VENEZUELAN OIL
Development and Chronology

VENEZUELAN OIL
Development and Chronology

ANÍBAL R. MARTÍNEZ

Graduate School of Administration of Hydrocarbons, Caracas Central University, Venezuela

ELSEVIER APPLIED SCIENCE
LONDON and NEW YORK

ELSEVIER SCIENCE PUBLISHERS LTD
Crown House, Linton Road, Barking, Essex IG11 8JU, England

Sole distributor in the USA and Canada
ELSEVIER SCIENCE PUBLISHING CO., INC.
655 Avenue of the Americas, New York, NY 10010, USA

WITH 18 TABLES AND 39 ILLUSTRATIONS

British Library Cataloguing in Publication Data

Martínez, Aníbal R.
Venezuelan oil: development and chronology
1. Venezuela. Petroleum industries, to 1988
I. Title
338.2′7282′0987

ISBN 1-85166-276-6

Library of Congress Cataloging-in-Publication Data

Martínez, Aníbal R., 1927–
Venezuelan oil: development and chronology/Aníbal R. Martínez.
p. cm.
Bibliography: p.
Includes index.
ISBN 1-85166-276-6
1. Petroleum industry and trade—Venezuela—History. I. Title.
HD9574.V285 1989
338.2′7282′0987—dc19

Printed by The Universities Press (Belfast) Ltd.

PREFACE

Venezuela has produced one tonne of crude oil for every seven used by man. With such abundant oil reserves as the Bolívar Coastal field in the west of the country and the Orinoco Belt field in the east, Venezuela will continue to be an important oil producer well into the twenty-first century.

Before the Europeans arrived in the Americas, the indigenous peoples were well aware of the substances occurring naturally on the surface. The Indians may not have realised their real worth, and took only what they needed from seeps for caulking boats, firming walls, and for medicinal and lighting purposes. The Spanish conquerors, and British and French buccaneers for that matter, profited in this knowledge.

From 1128, hydrocarbons belonged to the Crown and were not for private gain. The Royal Mining Ordinances allowed very little exploitation of the natural asphalt and very heavy crude oil. The Hispanic Republics inherited the reserves, and Codes and Permits to work the mines in Venezuela date back to 1854. The first company formed for the purpose of oil exploitation was Petrolia, which held sway for more than 60 years.

The 'big oil hunt' really began in the twentieth century with the blow-out of the Los Barrosos No. 2 well near the eastern shore of Lake Maracaibo on 14 December 1922.

The beginnings of the oil industry in Venezuela were as wild and disordered as elsewhere. In those early days, the unskilled owners of the resources were no match for foreign professionalism and cunning. Plentiful reserves of oil were extracted before the Venezuelans were expert enough to realise how they had been manipulated.

The 1950s marked the height of exploration, with the discovery of 200 oilfields and over 60% of proved reserves. In the 1960s, the State oil company, CVP, was created and then OPEC with other oil-exporting nations. Venezuela prepared to reserve her hydrocarbon industry and commerce.

The end of concessions on 31 December 1975 came about without conflict, with compensation paid to former concessionaires generously supplemented by so-called technology support and marketing contracts. Petróleos de Venezuela effectively assumed responsibilities for all industry operations. It also revived exploration, introduced research, gained sensible sale outlets, and later went on to internationalisation-downstream and soon, hopefully, upstream.

The Bolívar Coastal field is in its eighth decade, pilot programmes in the Orinoco Oil Belt have confirmed earlier expectations, and new discoveries include giant natural gasfields in the Margarita basin, the El Furrial structural trend in the Maturín basin, and the Guarumen prospect, southeast of Maracaibo.

From 1963, cumulative production had surpassed proved reserves. The picture began to change in the 1980s, however, when the operating subsidiaries of Petróleos de Venezuela showed that the 5-year evaluation plan in the Orinoco Oil Belt would bear fruit.

Petróleos de Venezuela entered the field of petrochemicals in November 1978 and, as scheduled, Pequiven was solvent within 7 years. In 1985, the Federal Executive gave Petróleos the responsibility for exploitation of the largest coal horizons so far measured in Venezuela, 70 km northwest of Maracaibo in the Guasare River valley; 2 years later, the first shipment left Venezuela. With the acquisition of a 50% share in eight refineries in Europe and the USA, Petróleos has assured outlets for one-third of its exports.

This book simply attempts to chronicle events as they happened. Chronologies are useful to scholars, teachers, executives, professionals and technicians alike, in fact anyone who needs to refer quickly to dates and events. Therefore, it is necessary that a chronology be accurate, and my efforts have been directed to this end. I have painstakingly checked and cross-checked my sources and, wherever possible, I have given the day and month of the event in question. Where this was not possible, I have merely listed the event at the beginning of the year or month, depending on the amount of information available.

I have endeavoured to cover all aspects of the Venezuelan petroleum industry's history – technical, legal, economic, social and political.

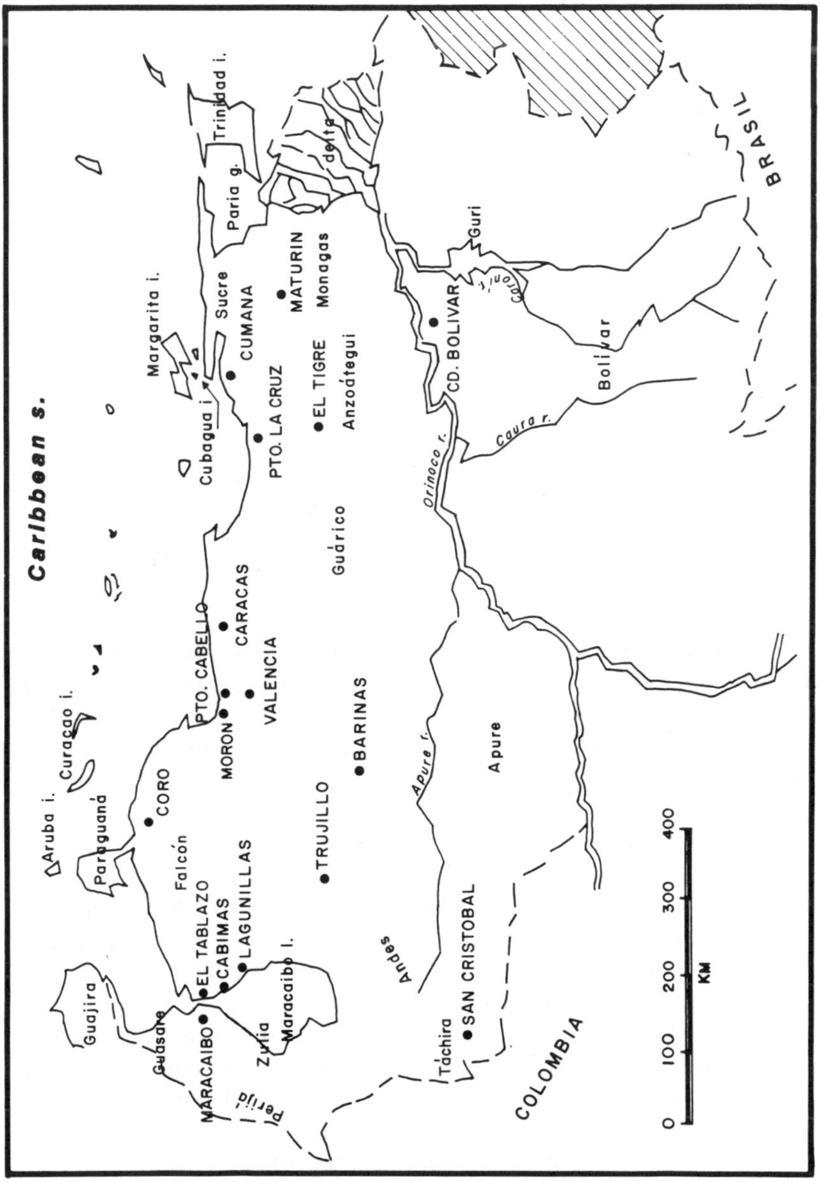

Fig. 1. Geographical index.

Certainly, because of the very nature of this book, events tend somewhat to be reduced to the same level. For example, the Hydrocarbons Law of 13 March 1943 was much more significant for the Venezuelan oil industry than that of 18 July 1928, and the discovery of deep reservoirs of Cretaceous production in the La Paz field was much more important than the discovery of the small Quiroz field. (However, the latter is of great personal value, for my first professional assignment was carried out there, scouting on Shell's QZ-1.)

New fields are discovered after a successful test on a petroleum-bearing section or after the completion date of a discovery well. Not all oilfields in Venezuela – approximately 470 at the end of 1987 – are listed, though this is not necessarily due to small reserves. The Bolívar Coastal field began as many separate units of production until, after many years of exploration, these units merged to form a whole – one single oilfield. I think it is quite erroneous to refer to 'Lagunillas' or 'Ceuta' as oilfields, for example.

Drilling in the Orinoco Belt field began in January 1936 with the testing of wildcat La Canoa No. 1 and, until the Seventh World Petroleum Congress in 1957, it was thought that the entire southern third of the Maturín basin was a 'tar belt'. However, characterisation of the fluids and meticulous probing have shown the field to be the largest crude oil deposit in the world, with six main producing areas.

Antecedents of Venezuelan oil legislation are found in the *Fuero Viejo de Castilla,* the *Royal Ordinances on Mining for the Colonies in the New World* and the *Quito Decree of Simón Bolívar.* The first laws were to do with mining in general, but soon they were specific to hydrocarbons. The amount of fiscal revenues raised from the exploitation of oil was a result of the interplay between the petroleum and income tax laws. A series of agreements between the Government and the oil companies were concluded on related matters, such as the calculation of royalty payments by means of reference to crude oils produced outside Venezuela. Legislation and contractual agreements with the oil workers' unions were handled with particular care, but in this chronology, I have not reported changing petroleum legislation in detail, for I thought it out of place.

Not all events have been included, for I had to apply strict criteria to the selection of events. I have tried to include those events significant to the development of the industry, those which were influential in shaping future policy, and those which precipitated further action. However, with special reference to more recent history, events may not have been included that, a few years hence, will show us to have been of the utmost significance – I can only interpret events as I see them.

The social and economic trends in Venezuela have greatly influenced events, as in other developing nations. Purely political matters are not mentioned, though I have included important national events and government policies are covered. The key figures in the development of Governmental policies emerge quite naturally and effortlessly. The policies of concessionaires are included on the rare occasions when they were self-evident or made public knowledge.

Since the 1940s, Venezuela has been active in international oil matters, and thus important developments are included. Venezuela, along with the other oil-exporting members of OPEC, are working together to create a unified petroleum policy, one that will safeguard their interests. However, due regard is being given to investors so that they secure a fair return on their capital, and to consumer countries so that they receive an efficient and economic supply of oil.

I hope I have covered adequately the events leading to the creation of the Latin American State Oil Companies' Mutual Assistance Association (ARPEL) and the Latin American Energy Organisation (OLADE). Also included are the international activities of the Petroleum Federation of Workers.

To a limited extent, I have also mentioned outstanding international events which have had a bearing on national developments, e.g. the expropriation decree of the Mexican President Cárdenas and the nationalisation programme carried out by Mossadegh. In particular, however, I have covered those actions of the USA and other nations who have posed difficulties for Venezuela, e.g. the US imports control programme.

The first reference in the literature to petroleum in Venezuela dates back to 1535, and the seep accounted for by Gonzalo Fernández de Oviedo y Valdés does indeed exist. In the first decade of the nineteenth century, Humboldt produced the first scientific report on the physiography and geology of Venezuela, including complete references, to naturally occurring hydrocarbons and thermal springs. The geological literature grew rapidly after the publication in 1841 of Colonel Codazzi's three-volume work on the physical and economic geography of Venezuela. German-speaking naturalists and geologists were in the main responsible, and the emphasis of their work, quite naturally, was on the asphalt deposits.

The twentieth century has seen the geological knowledge of the Venezuelan basins grow, though little of real importance was published early on except a large and inaccurate book by Liddle in 1928. At the Geological Congresses in 1937 and 1938, however, a better understanding of the geology of Venezuela was made public when concessionaire companies ended their policy of secrecy due to competition. The State oil company, CVP, submitted reports on the Gulf of Venezuela basin, the tar sands belt and the progress of exploration in the Guanarito area, to the Seventh World Petroleum Congress.

Six months after the first mining Code was promulgated, abrogating the *Ordenanzas* of Aranjuez, a concession to exploit natural asphalt deposits was granted to D.B. Hellyer in September 1854. The first crude oil concession was granted in 1865 by the constitutional President of Zulia State. The concessionaire could drill, produce and export petroleum and naphtha, 'or whatever the name of the oil which exists in the earth'. The concession was granted for 10 years, but it was cancelled after less than 12 months due to non-compliance with the provisions in the contract.

Ten years later, that is 20 years after the 'Drake' well adventure, Venezuela had its first commercial oil company – Petrolia. The importance of Petrolia is not merely historical, nor was the establishment of the company a romantic event devoid of consequences. Petrolia was truly Venezuelan, an integrated company that was able to compete favourably with foreign concerns for many years.

For many years I have enjoyed this undertaking. Even to risk prolixity, I have recorded the purely curious – the *Leona* field discovered by the *Tigre* wildcat – and the grotesque, together with an event's historical occurrence and overall importance. At the end, I came to see that history, as so often has been said, repeats itself. Some might feel that 1966 was the turning point for the Venezuelan oil industry, a year when reference prices were introduced, the income tax law was amended, CVP was strengthened and a number of actions were carried out by the State companies of OPEC Members. Others may feel that 1988 was the year, with the external debt problems, the new Orimulsion fuel, core flow testing, and general uncertainty within the industry.

I first began work on this chronology when studying at Stanford University. A considerable amount of my time was spent in the

Branner Library researching petroleum engineering and geology, wherever possible in relation to Venezuela. I soon acquired the habit of recording events which had occurred during the development of the Venezuelan oil industry. I even earned two credits for a rough, preliminary compilation of 289 events, entitled *Chronology – Venezuelan Oil.*

This habit of note-taking has stayed with me and, over the years, I have continued recording important events relating to the oil industry in Venezuela – whether from course work, scientific publications, magazines or other documents.

A list of acknowledgements to all those people who have helped me prepare, expand or correct this chronology would be too unwieldy. However, let me say that any errors, misinterpretations or omissions are solely my responsibility.

I dedicate this book to Mary; Miriam, Elizabeth, Carolina, Yvette and Gisela; Carlos and Juan José; Alex, Thomas, Juan José and Daniel; and, who knows, to those of forthcoming events.

ANÍBAL R. MARTÍNEZ

LIST OF ILLUSTRATIONS

LIST OF TABLES

ABBREVIATIONS

Bs	bolívars
cm	centimetres
gal/day	gallons per day
gal/month	gallons per month
ha	hectare
km	kilometres
km^2	square kilometres
litre/day	litres per day
m	metres
m^3	cubic metres
m^3/day	cubic metres per day
ton/year	tonnes per year

ARPEL	State Oil Companies of Latin America
BCF	Bolívar Coastal field
CEE	European Common Market
GIPLACEP	Informal Group of Latin American Oil Producers
LAFTA	Latin American Common Market
OLADE	Latin America Energy Organisation
OPEC	Organisation of Petroleum Exporting Countries

CONTENTS

The Chronology

THE CHRONOLOGY

In early pre-Discovery days the Indians were well aware of the conspicuous natural surface occurrences of hydrocarbons which exist in Venezuela. Of course, they could not realise that one day the substance they used for their medicines and illuminations was to become known as oil and could swing the balance of world economics and politics. The Indians used the word *mene,* which in fact is still in use today for the seepages they had discovered. Together with these seepages, the Indians found asphalt, which they used for caulking their canoes and impregnating the sails of their boats, and pitch, which they used for lining their hand-woven baskets. They hunted deer and other wild animals in and around the seepages (turned into excellent trapping grounds by the heat of the sun). In what is today Trujillo State, the Indians collected lighter oils from along the small creeks near the seepages by impregnating blankets and then wringing them out.

1128

The *Fuero Viejo de Castilla* declares that no one could work mines in the 'land property of the King'.

1389

King John I of Spain gives 'to the people' the right to explore and look for mines, though they are obliged to pay two-thirds of any finds in tribute.

1499–1528

The Spanish conquerors are greatly impressed with the natural occurrences of hydrocarbons in Venezuela. They learn from the Indian *piaches* to use the substances for medicinal purposes; they also use them for caulking their ships, for illumination and for treating their weapons.

1526

December 9
King Charles I of Spain issues an edict on mining, declaring ownership on mines discovered in America.

1532

December 10
King Charles I of Spain gives permission to the islanders in Cubagua to use the fountain of natural bitumen for medicinal purposes.

1535

September 30
Gonzalo Fernández de Oviedo y Valdés, in his *General and Natural History of the Indies* (*Historia General y Natural de las Indias, Islas y Tierra-Firme del Mar Oceano*), is first to mention Venezuelan oil seepages. He reports an 'abundant' seep at the western end of Cubagua island. The 'nectar from Cubagua' is very useful for many things and for various sicknesses, in particular gout and others due to coolness, he said. Some had called the substance *stercus demonis*.

1536

September 3
The Queen of Spain, Joanna, in a letter from Valladolid to 'the officials who send of the oil petroleum' (*azeite petrolio*) from Nueva Cádiz, Cubagua, orders that 'as much of it as possible be sent to me in all ships coming from said Island'. The Queen mentions 'a fountain' of oil in Cubagua and that all the amounts already sent 'had seemed to be of good'.

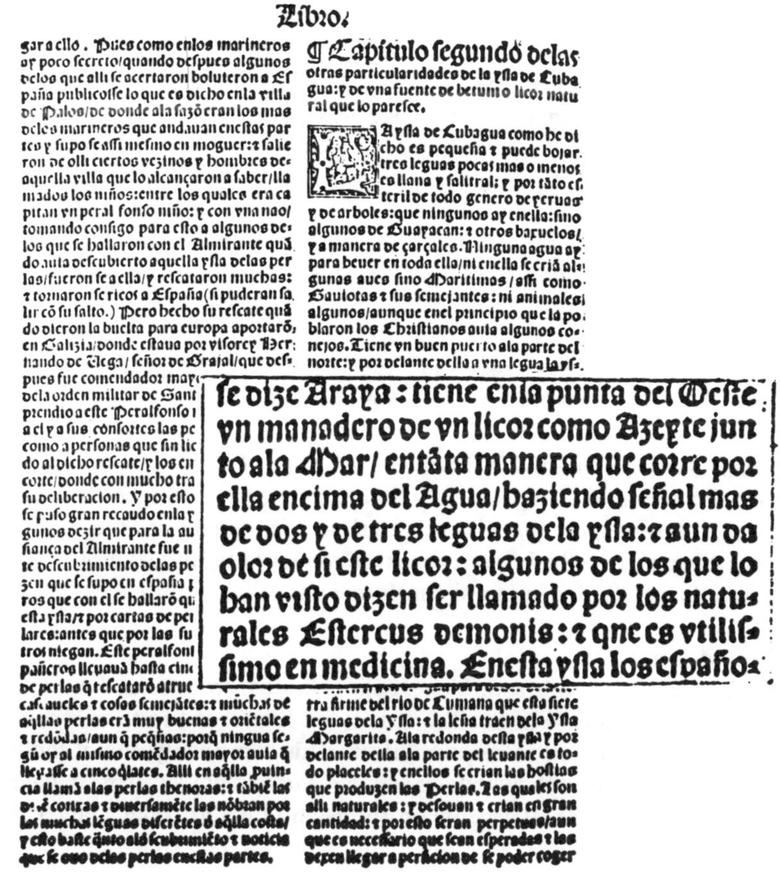

Libro.

gar a ello. Pues como enlos marineros ay poco secreto/quando despues algunos delos que alli se acertaron boluieron a España publicosse lo que es dicho enla villa de Palos/de donde ala sazō eran los mas delos marineros que andauan enestas partes y supo se assi mesmo en moguer: e salieron de alli ciertos vezinos y hombres de aquella villa que lo alcançaron a saber/llamados los niños: entre los quales era capitan vn peralfonso niño: y con vna nao/ tomando consigo para esto a algunos delos que se hallaron con el Almirante quādo auia descubierto aquella ysla delas perlas/fueron se a ella/y rescataron muchas: e tornaron se ricos a España (si pudieran salir cō su salto.) Pero hecho su rescate quādo dieron la buelta para europa aportarō, en Galizia/donde estaua por visorey Hernando de Vega/ señor de Grajal/que despues fue comendador mayor dela orden militar de Sant... prendio a este Peralfonso ... a el y a sus consortes las pe... como a personas que sin lic... do al dicho rescate/y los en... corte/donde con mucho tra... su deliberacion. Y por esto se puso gran recaudo enla y... gunos dezir que para la au... sencia del Almirante fue n... te descubrimiento delas pe... zen que se supo en españa ... ros que con el se hallarō q... esta ysla/e por cartas de pe... lares: antes que por las su... tros niegan. Este peralfons... pañeros lleuauā hasta cin... de perlas q̄ rescatarō a true... cascaueles e cosas semejātes: e muchas de aquellas perlas erā muy buenas e orietales e redōdas/aun q̄ pequeñas: porq̄ ninguna segū oy al mismo comēdador mayor auia q̄ llegasse a cinco qlates. Alli en aqlla punta llamā alas perlas tebenoras: e tābiē las ... contras e diuersamēte las nōbran por las muchas leguas diferētes d aqlla costa/ y esto baste quāto alo descubrimiēto e noticia que se ouo delas perlas en estas partes.

¶ Capitulo segundo delas otras particularidades de la ysla de Cubagua: y de vna fuente de betum o licor natural que lo paresce.

La ysla de Cubagua como he dicho es pequeña e puede bojar tres leguas pocas mas o menos es llana y salitral: y por tāto esteril de todo genero de yeruas y de arboles: que ningunos ay enella: sino algunos de Guayacan: e otros baxuelos/ y a manera de çarçales. Ninguna agua ay para beuer en toda ella/ni enella se criā algunas aues sino Maritimas/assi como Gauiotas e sus semejantes: ni animales algunos/aunque enel principio que la poblaron los Christianos auia algunos conejos. Tiene vn buen puerto ala parte del norte: y por delante della a vna legua la ysla se dize Araya: tiene enla punta del Oeste vn manadero de vn licor como Azeyte junto ala Mar/ entāta manera que corre por ella encima del Agua/haziendo señal mas de dos y de tres leguas dela ysla: e aun da olor de si este licor: algunos de los que lo han visto dizen ser llamado por los naturales Stercus demonis: e que es vtilissimo en medicina. Enesta ysla los españo... tra firme del rio de Cumana que esta siete leguas dela Ysla: e la leña traen dela Ysla Margarita. Ala redonda desta ysla y por delante della ala parte del leuante es todo placeles: y enellos se crian las hostias que produzen las Perlas. Las quales son alli naturales: y desouan e crian en gran cantidad: e por esto seran perpetuas/aun que es necessario que sean esperadas e las dexen llegar a perfecion de se poder coger

Fig. 2. First reference in the literature to Venezuelan petroleum: a seep in the sea, west of Cubagua Island. In Gonzalo Fernández de Oviedo y Valdés, *General and Natural History of the Indies,* Seville, 30 September 1535.

1539

April 30

One barrel of Venezuelan oil is sent to Spain in the *nao* 'Santa Cruz', Mestre Francisco Rodríguez de Covarrubia and Captain Bernardino de Fuentes, by Don Francisco de Castellanos, Treasurer of Nueva Cádiz, 'to alleviate the gout of Emperor Charles V'. This is the first documented shipment of petroleum from the Western Hemisphere.

October 18
Following the arrival in Spain of the *nao* 'Santa Cruz' the Queen sends a letter from Madrid to the House of Trade (*Casa de Contratación*) in Seville requesting that the 'barrel of oil petroleum' be forwarded to her.

October 31
The barrel of oil is sent by the *Casa de Contratación* to the Queen in Madrid in the custody of Alfonso García, a muleteer (*arriero*) from Cuerva.

1540

In the Second Part of his *General and Natural History of the Indies*, Gonzalo Fernández de Oviedo y Valdés, reports on the occurrence of many oil seepages (*ojos ò manantiales de betun*) in the Gulf of Venezuela area, 'which the Indians call *mene*'. He says some seeps are 'one-quarter of a league in circumference'.

December 14
Another barrel of oil from Venezuela is known to have arrived at the *Casa de Contratación* in Seville. It is to be forwarded to the Queen 'in the custody of person of responsibility'. This is probably one of the last oil shipments through Cubagua, as an earthquake destroyed Nueva Cádiz and all other settlements on the Island on Christmas Day, 1541.

1551

December 17
By Royal Declaration it is established that, like the Spaniards, Indians can discover, own and work mines.

1552

Francisco López de Gomara, in his *General History of the Indies* (*Historia General de las Indias*), repeating Fernández de Oviedo y Valdés, refers to a seepage in Cubagua and the corresponding stain in the neighbouring sea.

1559

Mining Ordinances of King Phillip II in Valladolid annul all previous rights and incorporate all mines to the royal patrimony.

1579

June 15
Mayor Gaspar de Párraga and Rodrigo de Argüelles of Maracaibo report of an oil seepage near the city of Nueva Zamora (Maracaibo) and four more 'in the province'. They also give a detailed description of how the oil is used locally.

1584

August 22
King Phillip II of Spain sanctions the *San Lorenzo* Ordinances, a compilation of mining regulations in force.

1589

Juan de Castellanos, in his *Elegies* (*Elegías de Varones Ilustres de Indias*), refers to a 'fountain' seepage in Cubagua. He says that 'proven and excellent liquor' is commonly used for medicinal purposes. When the sea currents are strong, oil stains extend for 'three leagues' in the sea.

1602

The *Leyes de Indias* authorise the Royal Governors to apply the laws of Spain throughout the American colonies.

1615

Antonio Herrera refers to the 'fountain of smelly medicinal liquor' in Cubagua in his *General History* (*Historia General de los hechos de los*

Fig. 3. Guanoco natural asphalt lake after the rain.

Castellanos en las islas i Tierra Firme del Mar Oceano).

1641–1678

French and English buccaneers, who roamed the Caribbean in their anti-Spanish piratical adventures, enter Lake Maracaibo, chiefly by force, to repair their sea-battered ships with pitch from the many seepages bordering the lake. Gerard, Jackson (twice), Montbars, Nau, Morgan and Granmont sacked Maracaibo or Gibraltar (at the south end of the lake); Granmont went as far inland as Trujillo.

1680

May 18
King Charles II of Spain publishes a *Recopilación de Indias* containing all previous legal references to mines and mining, and orders that the San Lorenzo Ordinances be applied to America.

1741

Joseph Gumilla, in his *Natural, Civil and Geographical History of the Orinoco Region* (*Historia Natural, Civil y Geográfica de las naciones situadas en las riberas del Río Orinoco*), reports on the use of pitch from Trinidad, mostly for medicinal purposes.

1766

Reports from Cariaco and Aguas Calientes refer to gases and hot water flows due to an earthquake in Cumaná.

1783

May 22
King Charles III of Spain dictates the Mining Ordinances for New Spain (*Ordenanzas de Minería para la Nueva España*) at Aranjuez. Mines are declared the property of the Royal Crown; the King can grant mining concessions. Disposition 22, Sixth Title, specifies 'any other fossils, be they perfect or semi-minerals, bitumens, or juices of the earth'.

1784

April 27
By Royal Resolution and Royal *Cédula*, the *Ordenanzas de Minería* of Aranjuez are applied to the Intendency of Venezuela.

1799

July 16
The naturalists Alexander von Humboldt and Aimé Bonpland arrive in Venezuela (Cumaná).

1806

F. Depons, in his *Voyage à la partie orientale de la Terre Firme, dans l'Amérique Méridionale, fait pendant les années 1801, 1802, 1803 et 1804*, refers to oil seeps to the northeast of Lake Maracaibo, near a place called *Mena* (?), and concludes that vapours emanating from them are the cause of the 'Catatumbo lightning' (which is a purely atmospherical phenomenon).

1811

Nicholas Nugent, an honorary member of the London Geological Society, reports that great quantities of woody and vegetable bodies rolled down by the vast Orinoco river accumulated in particular places, underwent transformations and chemical changes, and formed bitumen.

1813

J.J. Dauxion Lavaysse, in his *Trip to Trinidad, Tobago and Margarita Islands* (*Vogage aux Iles de Trinidad, de Tabago, de la Marguerite et dans diverses parties de Vénézuéla, dans l'Amérique Méridionale*), refers to seeps near Cumaná and Barcelona.

1814

Alexander von Humboldt gives the first serious account of the asphalt deposits in Venezuela in his *Travels to the Equinoctial Regions of America* (*Reise in die Aequinoctial-Gegenden des neuen Continents in den Jahren 1799, 1800, 1801, 1802, 1803 und 1804*). He relates the deposits to the causes which produce earthquakes and eruptions of lava in northern South America. He also describes how the natives living near seepages use the pitch and the asphalt. Humboldt gives a list of the natural occurrences of hydrocarbon and hot springs from Trinidad to Maracaibo.

1816

In *An Exposition of the Commerce of South America,* Manuel Torres indicates that Spanish America possesses 'on the banks of its navigable rivers, abundant stores of bitumen, coal, and other fossil substances'.

1823

February 11
In a report to the French Academy of Sciences, Jean Baptiste Boussingault describes the Onoto and Mariara hot springs, near Maracay.

1825

Samples of light oil from a seepage reportedly located between Escuque and Betijoque are sent to the United Kingdom, France and the United States.

1829

October 24
Simón Bolívar dictates *Reglamento sobre Minas* decree in Quito, reiterating ownership by the Republic of 'mines of all kinds'.

1830

A group from El Moján (near Maracaibo) explores the Sierra de Perijá area (Socuy river). A burning gas seepage (believed by them to be a volcano) frightens them.

Charles Lyell reports in his treatise on geology that organic substances carried for centuries by the Orinoco river could accumulate in particular places and be converted into petroleum or asphalt, as in the island of Trinidad.

1832

April 29
The Venezuelan Congress ratifies the Bolívar decree dictated in Quito, on national ownership of 'mines of all kinds'. The applicability of the *Ordenanzas de Minería* of Aranjuez is also accepted.

1839

October 3
José María Vargas reports to the Ministry of Finance on a sample of asphalt sent to him. He refers to its occurrence in Venezuelan soil and to the fact that he already had 'a bottle of this oil brought from the Trujillo Province'. He proposes further investigation to determine the real extent, form of occurrence and depth of the deposit, 'daring to consider' that the State might even grant concessions for exploitation of the resource. The sample had been collected at Pedernales, Lower Orinoco canton.

1841

Agustín Codazzi refers to 'fountains of oil' near Trujillo and Cumaná, and to asphalt deposits in Mérida, Coro and Maracaibo, in his *Summary of the Geography of Venezuela.*

1849

May 3
A small quake, felt in Maracaibo, is associated with the Perijá 'volcano'.

1850

First summary of geology of central and eastern Venezuela is published by Hermann Karsten in the *Journal of the German Geological Society* (*Zeit. der Deutschen geol. Ges.*). The accompanying map shows the location of numerous hot springs.

1851

Hermann Karsten refers in an article on Venezuelan geology, published in the *Karsten Archiv* (Berlin), to an oil seep between Escuque and Betijoque (*Bottijoque*).

1852

September 20
Hermann Karsten reports from Barranquilla (on 20 September) that petroleum seepages can be found in many places all around Lake Maracaibo.

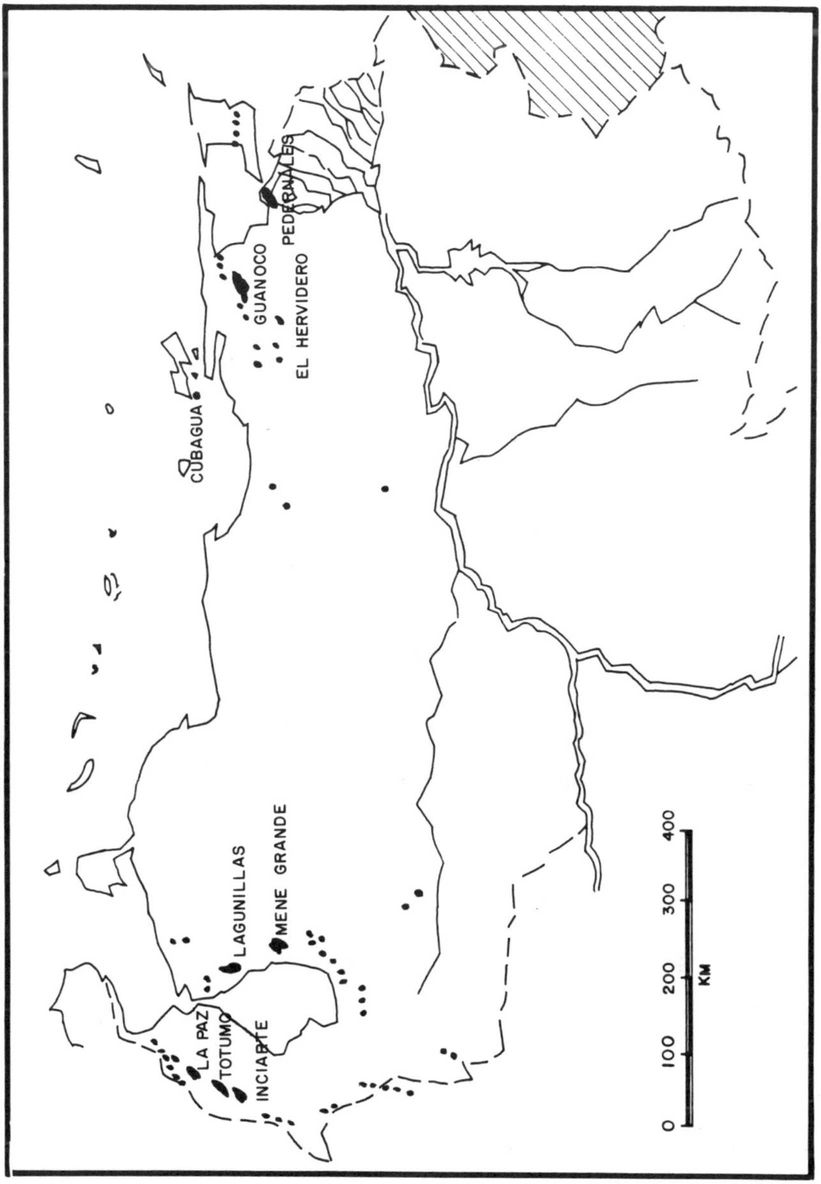
LA PAZ
TOTUMO
INCIARTE
LAGUNILLAS
MENE GRANDE
CUBAGUA
GUANOCO
EL HERVIDERO
PEDERNALES
0
100
200
300
400
KM

Fig. 4. Seeps.

1854

May 15
First Venezuelan Mining Code (*Código de Minas*) in eight laws is enacted, abrogating *Ordenanzas de Minería de Aranjuez.*

September 6
A concession for the exploitation of asphalt is granted to D.B. Hellyer.

1855

The area west of Lagunillas (Zulia) is explored. Burning gas and oil seepages are related to volcanic activity.

January 4
Presidential Decree of José Gregorio Monagas contains Regulations for 1854 Mining Code. Article 2 establishes need of concessions to exploit combustible deposits, which are the property of the State.

1857

May 6
A Development Board (*Junta de Hacienda*) asks permission from the President of Zulia to exploit the asphalt in the State and that no contracts be given to the private persons which have asked for them.

1858

The oil industry sees its first production. James Miller Williams begins producing oil from shallow holes dug at Oil Springs, Ontario, Canada.

1859

August 27
'Colonel' Edwin Drake completes the first oil well in the United States.

1860

William Barnsdall and William Abbot build the first oil refinery, near the Drake well.

May 16
In a report read to the Geological Society of London, G.P. Wall refers to mineral pitch 'extensively diffused in the province of Maturín . . . and in still larger quantities near the Gulf of Maracaybo' and to mud volcanoes near Maturín (locations shown in accompanying map).

1864

April 13
The Federal Constitution transfers to State property all petroleum and combustibles.

1865

August 24
General Jorge Sutherlan, Constitutional President of Zulia State, grants the first oil concession in Venezuela to Camilo Ferrand, of the USA. The lease covered the State for a duration of 10 years; Ferrand could drill, produce and export (*taladrar, sacar y exportar*) petroleum or naphtha – 'or whatever the name is of the oil which exists in the earth'. A royalty of 5 pesos per ton was fixed, although the State could request payment of 20 000 pesos at agreed intervals. The State did not exempt Ferrand from the payment of duties for importing machinery needed for the operations.

1866

February 2
The legislature of Nueva Andalucía State (now Sucre and Monagas) grants an oil concession to Manuel Olavarría. The lease covers the whole State for 25 years; a royalty of 17% is fixed and Sr. Olavarría has to carry out certain improvements to the cities of Cumaná and Maturín at his own expense.

1867

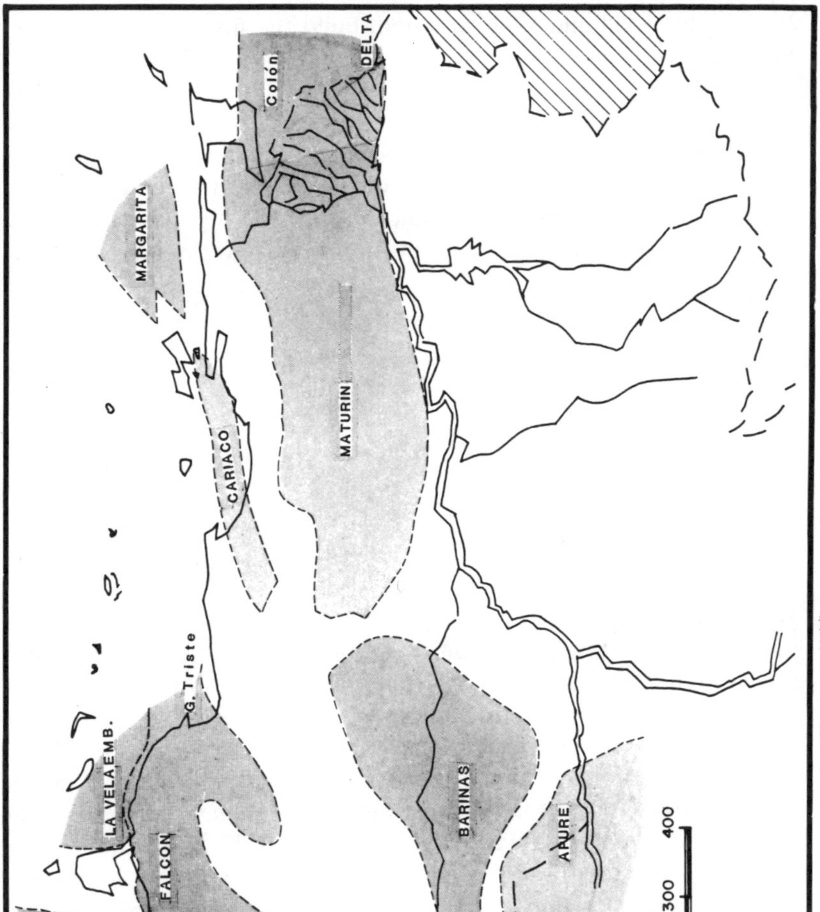

Fig. 5. Sedimentary basins.

February 23
The Ferrand concession lapses.

December 19
The Constitutional Assembly of Trujillo State grants a concession near Escuque to Pascual Casanova for a duration of 20 years.

1867

May
Adolf Ernst founds the Caracas Society for Natural Physical Sciences.

May 20
The first Civil Code establishes that ownership of mines is not necessarily that of the landowner.

1869

April 12–14
Arístides Rojas publishes a three-article series in *La Opinión Nacional,* referring to the line of hot springs, asphalt deposits and seepages in northeastern Venezuela and Trinidad (Chapapote – Mexican word for *mene*-, Aragua, Chaguaramal, Araya, mouth of Guarapiche).

1870

Arístides Rojas, in a book on the geography of Venezuela, gives numerous references to asphalt deposits and petroleum seepages.

1873

A report in the bulletin of the Ministry of Development refers to asphalt and petroleum 'mines' in Trujillo (near Escuque), Falcón and Araya (near Cariaco Gulf).

1874

Adolf Ernst establishes a chair of Natural History at the Central University of Caracas.

1875

In a book published in Paris (*Venezuela Pintoresca e Ilustrada*), Miguel Tejera refers to two known 'mines' of petroleum in Trujillo (Escuque), and many more in Falcón and in Araya. He also refers to abundant asphalt deposits: 'mines' near Gibraltar, Altagracia (Zulia), Guanoco and 'inexhaustible' deposits in Mérida.

May 18
Cúcuta (Colombia) earthquake. In Venezuela, some 30 km to the southeast, in the coffee plantation of Manuel Antonio Pulido, sticky oil starts oozing out of a large fracture. He commences the promotion of a company to exploit the deposit.

1876

February 8
In an official (anonymous) report to the President of the Zulia State, 'asphalt and *mene* mines' in the Misoa, Cabimas and Perijá areas are discussed.

June 30
Eng. Gen. Wenceslao Briceño Méndez submits to the President of Zulia a very accurate and complete report on the petroleum and asphalt deposits in the Maracaibo basin: west of Maracaibo, Perijá, Río de Oro, Tarra, eastern shores of the lake. He refers to an oil seep in Tarra producing 5760 gal/day.

1878

September 3
M.A. Pulido secures from the government of the Gran Estado de Los Andes (Táchira) exclusive rights to exploit a 100-hectare parcel 15 km west of San Cristóbal, near Rubio.

October 12
M.A. Pulido and associates establish the Compañía Nacional Minera Petrolia del Táchira.

1879

March 10
Pedro Rafael Rincones, one of the partners of M.A. Pulido, travels to Pennsylvania to study the oil industry. P.R. Rincones buys a drilling rig and refinery equipment for Petrolia operations.

1880

The Star drilling rig is shipped from Pennsylvania to La Alquitrana, the concession site of Petrolia. In the meantime, shallow pits had been dug and petroleum scooped out with buckets or, Indian fashion, wrung out of impregnated cloths.

1881

April 27
The new Constitution declares that although States are owners, the administration of mines rests with the Federal Government.

1882

May
A teapot-like still with a 2000-litre daily capacity is set up by Petrolia in La Alquitrana, not very far from the producing area.

July 31
Petrolia is legally constituted. Carlos González Bona is appointed President.

End of October
Petrolia abandons first well.

1883

March 3
First Law – Decree on mining referring to the 1881 Constitution is published.

April
Petrolia completes the ‘Eureka’ well at a total depth of 60 m with an output of 230 litres/day, discovering the La Alquitrana field (*Maracaibo basin, 18 km southwest of San Cristóbal*).

May 7
A concession to exploit the Guanoco asphalt lake is granted to Horacio R. Hamilton and Jorge A. Philips.

May 8
The Minister of Development contracts refining of all crude oil produced in Venezuela mines with Roberto García.

September 22
Cristóforo Dacovich is authorised to explore petroleum mines he had denounced in Unión site, Escuque, Trujillo State.

November 15
Second Law – Decree on mining referring to the 1881 Constitution is issued. The duration of the mining concession is fixed at no more than 99 years, no less than 50 years.

1884

A concession to exploit oil and other minerals is granted to Sixto González, covering part of the Guárico State.

February 11
William L. Lay of New York reports to Manuel Cadenas-Delgado on petroleum and mineral wax lands at headwaters of Caús River, near Escuque.

March 25
A concession for 99 years to exploit oil and asphalt, covering part of the Zulia State, is granted to José Andrade of Maracaibo.

April 8
The Federal Executive validates the Petrolia concession.

C. Dacovich is granted a concession of 12 petroleum mines for 20 years in Caús, Escuque, Trujillo State.

May
Production of La Alquitrana field is 6 m^3. First products exported by Petrolia to Colombia.

July 3
M. Cadenas-Delgado is authorized to explore petroleum and ozoquerite mines in Betijoque and Escuque.

August 9
Manuel Hernández-López signs a contract to explore for asphalt and petroleum in the Paraguaná peninsula for 99 years.

October
Wilhelm Sievers arrives in Venezuela.

December 31
Consumption of illuminating oil in Venezuela is 2000 boxes of 2 cans, or 328 000 litres/day.

1885

February 27
C. Dacovich and associates establish a company to exploit the Colombio petroleum mines near Escuque.

May 23
A new Mining Code (*Código de Minas*) and Ordinance is issued.

August 13
C. Dacovich is granted an exclusive concession to refine oil in the Andes and Falcón States for 20 years.

October 24
H.R. Hamilton offers his concession to exploit the Guanoco asphalt lake to the New York and Bermúdez subsidiary of General Asphalt.

December 9
Transfer of Guanoco Hamilton concession is not approved by the Government.

1886

New York and Bermúdez, a subsidiary of General Asphalt, succeeds in acquiring the Hamilton concession to develop the asphalt lake near Guanoco.

October
The El Salvador well, newly completed in the La Alquitrana field, produces 200 litres/day.

1887

May 30
A short six article Mining Code is passed. For the first time, reference is made to mineral substances – coal, asphalt, bitumen.

August 3
Regulations to the Mining Code (*Decreto Reglamentario*) are enacted. The long decree defines for the first time 'soil' and 'subsoil'.

1888

W. Sievers lists and describes 57 hot springs sites in Venezuela. He also mentions the oil-producing and -refining operations in the La Alquitrana field.

January 20
The Minister of Development signs a contract with Aníbal Domínici, representative of the Venezuelan Oil Company, for the exclusive rights to exploit, refine and export crude oil and products in Venezuela.

May 12
Charles Bullman writes in the *Engineering and Mining Journal* (New York) about the large asphalt deposits discovered in the delta of the Orinoco. He also reports that 'petroleum of superior quality is found in Tachira'.

October 24
First continuous electricity service in Maracaibo.

1889

September 8
A. Ernst writes about the geological evolution of Lake Maracaibo and nearby oil occurrences.

1890

Graham Company of Trinidad drills four shallow wells at La Brea, Capure Island, near Pedernales. The project is later abandoned.

1891

June 26
A new Mining Code (*Código de Minas*) is issued.

July 4
The first issue of the *Bulletin of Public Resources* (*Boletín de la Riqueza Pública*) includes a table showing asphalt deposits in many localities in Venezuela, also listing a 'very active oil seep' in El Infierno (Zulia) and a Pedernales asphalt deposit 'under exploitation'.

July 8
Contract with A. Domínici is declared null.

September 12
Regulations on the latest Mining Code are published.

September 19
Bulletin of Public Resources reports 4316 kg of asphalt were exported during 1890.

October 10
Ramón Alvarez and Fernando Gómez receive title to a petroleum mine discovered in Yaguaraparo, near Cumaná.

1892

January 9
The *Bulletin of Public Resources* report Andrés E. Level owns asphalt mines in Orinoco delta area 'considered excellent by Dr José María Vargas'.

1893

March 29
Another *Código de Minas* is passed. The word 'petroleum' is used for the first time.

1894

May 1
C. Dacovich is granted a petroleum mine in Encontrados.

July 15
An article in *El Cojo Ilustrado* reports gas and electricity displaces usage of petroleum products for illumination in Caracas. Predictions are attempted concerning the development of the oil industry in Venezuela.

1895

E. Fortin reports to the Geographical Society (Paris) on the Guanoco asphalt lake.

1896

H. Eggers publishes a paper on the asphalt deposits of the Lake Maracaibo area.

1897

The first hydroelectric plant is inaugurated in El Encantado, on the Guaire river, near Caracas.

June 8
Antonio Bianchi and others receive the concession of 'an asphalt mine' in the Guanoco area, which is immediately opposed by the New York and Bermúdez Company.

1898

Wilhelm Sievers reports on the trips made during March of 1892 and 1893 by chem. Richard Ludwig to El Hervidero mud volcano, southeast of Maturín.

Clifford Richardson makes a detailed report on the 'Bermudez' asphalt lake (Guanoco).

1899

October 22
General Cipriano Castro occupies Caracas and takes over power. During his administration, the first important concessions are granted.

1900

The Val de Travers Asphalt Paving Co., an English concern, begins exploiting asphalt in the Pedernales area. A small asphalt refinery was erected.

A concession is granted to the United States and Venezuela Co. (Uvalde Asphalt Paving Co.), to exploit the Inciarte asphalt lake, Zulia State. The company operated for 4 years; a small asphalt refinery was built at Carrasquero. In 1905, the Government took back the concession.

1901

New York and Bermúdez begins exploitation of the asphalt deposits in the Guanoco lake. The mineral is exported to Brazil.

An Italian engineer, E. Cortese, reports on asphalt deposits and oil seeps in Venezuela at Guanoco and Guariquen, Putucual east end of Cariaco gulf, Manicuare, Maracaibo area and 'near Puerto Cabello'.

Lilia, a novel by Ramón Ayala, makes reference to Venezuelan oil for the first time.

1903

L. Hirzel writes about the petroleum and asphalt deposits in the delta of the Orinoco river (Pedernales, Plata island).

1904

January 23
A new Mining Code (*Código de Minas*) is issued, abrogating the Code of 1893.

January 28
New York and Bermúdez succeeds in its claim against A. Bianchi and others for the Guanoco concession.

May 16
A 1214-hectare concession is granted to Andrés Espina and others in Perijá and Maracaibo areas.

A contract for the exploitation of asphalt mines in Perijá and Maracaibo is concluded between the Federal Executive and Andrés Valbuena and Federico Bohórquez.

June 21
Regulations are issued on the acquisition and exploitation of asphalt, petroleum, bituminous tar and similar substances.

September 22
The National Government accuses the New York and Bermúdez Co. of cooperation with General Manuel Antonio Matos in the 'La Libertadora' rebellion.

1905

In London, C. Richardson publishes a report on Venezuelan asphalts ('Bermudez' and 'Maracaibo'), including their chemical composition, characteristics and location maps.

August 14
President Castro issues a Mining Law (*Regimen de la Ley de Minas*) under which many of the famous petroleum concessions of Venezuela were to be granted. An annual surface tax of $0.60 per hectare was established, plus a royalty of $1.20 per ton; concessions could be granted for 50 years and exploitation had to start within 4 years. Disputes were subject only to the Venezuelan law and courts.

November 12
Domingo Antonio Coronil receives a mining concession in Punceres, near Maturín.

November 16
Francisco Hernández receives a concession for exploitation of oil in Piar District, Sucre State.

December 16
A 50-year lease is given to Eduardo Echenagucia García covering the Zulia State for the exploitation of oil. The concession was later annulled because work did not start within the agreed time limit.

1906

February 23
The Regulations to the 1905 Law are issued. The executive has the absolute power to grant and administer all concessions without the consent of the National Congress.

March 26
The Federal Court (*Corte Federal y de Casación*) recognises the *Ordenanzas de Minería* of Aranjuez had indeed applied legally to Venezuela.

1907

January 31
A concession is granted for 50 years to Andreś Jorge Vigas in the Río de Oro area. It is later transferred to Colón Development.

February 28
Antonio Aranguren receives a special concession to exploit asphalt in the Bolívar District, Zulia State; the concession also covers Maracaibo District.

March 18
A concession is granted to Francisco Jiménez Arráiz for the exploitation of asphalt, pitch and tar in the Acosta and Zamora Districts of Falcón State and the Silva District of Lara State.

July 3
The Jiménez Arráiz concession in Falcón and Lara is amended to also include the exploitation of oil. The lease was later transferred to North Venezuelan Petroleum.

July 22
General Bernabé Planas is granted a concession for the exploitation of asphalt, bitumen, pitch and petroleum in the Buchivacoa district of Falcón State. It is later transferred to British Controlled Oilfields.

1908

August 1
Rafael Gutieri receives a concession for the exploitation of asphalt, bitumen, pitch and petroleum in the Puerto Cabello District of Carabobo State.

August 12
New York and Burmúdez must pay Bs24 million in legal proceedings for the annulment of the former Hamilton concession in Guanoco asphalt lake.

1908

C. Richardson publishes 'The Modern Asphalt Pavement' (London; pp. 178–92 are devoted to Venezuela). He gives detailed data on the 'Bermudez', 'Maracaibo', Inciarte and La Paz asphalts, as well as general information on the exploitation of the deposits.

January 3
First documented Venezuelan oil statistics. Manager C. González-Bona reports production for January 1905 to November 1907 in the La Alquitrana field at 990 barrels of 40 gal.

December 19
Less than a month after the departure of President Castro to Europe, General Juan Vicente Gómez seizes power.

1909

G. Delgado Palacios reports on the various types of naturally occurring hydrocarbons in Venezuela.

March 9
Diplomatic relations with the USA are resumed, following the 'La Libertadora' rebellion and acceptance of the sentence imposed on New York and Bermúdez.

April 12
Eloy Escobar Llamozas receives a 25-year concession to exploit asphalt in the Maturín area.

August 16
A greatly improved Mining Code (*Código de Minas*) is issued.

December 4
A concession covering the Inciarte asphalt lake is granted to E. Stanley Simmons.

December 10
A concession covering 27 million ha is granted to John Allen Tregelles, representative of a British company. It has for the first time a provision promoting local refining of oil produced and another permitting the expropriation for right of way.

1910

Pedro José Rojas reports asphalt deposits in Mene Grande area are harmful to farming.

L.V. Dalton, in a paper published in London on the geology of Venezuela, refers to the petroleum seepages and oil possibilities in the country.

June 29
A new Mining Code (*Código de Minas*) is enacted.

July 14
Rafael Max Valladares, attorney for General Asphalt, receives a petroleum concession covering the Benítez District of Sucre State (Guanoco pitch lake included). In 1912 the first oil well in eastern Venezuela is drilled under this lease by New York and Bermúdez.

August 11
J.A. Tregelles and others study the La Alquitrana field.

1911

February 11
The Inciarte asphalt lake in Zulia State is leased for exploitation to General José María García.

July
One well is drilled near Cumaná by Venezuelan Oil Fields Exploration in the Tregelles concession.

September
Ralph Arnold and his associates start one of the most ambitious exploration programmes in the world, the systematic and scientific study of the oil possibilities of General Asphalt holdings in Venezuela (an area of more than 50 million ha).

November
Caribbean Petroleum is incorporated in New Jersey by General Asphalt, to hold the concession being negotiated in Venezuela.

December
The Tregelles concession is taken back at the end of the prospecting period.

1912

A book on *Venezuela* is published in London by L.V. Dalton, containing references to the oil and asphalt industry throughout. Dalton reports the 'desultory and ill-advised' attempt to develop asphalt production at Pedernales and the satisfactory exploitation of oil by Petrolia. He remarks that Venezuela should be considered one of the important oil-producing countries of the world.

January 2
R.M. Valladares receives a concession covering roughly the same area granted to J. A. Tregelles, that is, all the Anzoátegui, Carabobo, Táchira, Monagas, Mérida, Lara, Trujillo and Yaracuy States, the Federal Territory of Delta Amacuro, plus parts of the States of Zulia, Falcón and Sucre.

Caribbean Petroleum registered in Philadelphia.

January 4
R.M. Valladares transfers his concession to Caribbean Petroleum.

January 6
The US Legation sends to the State Department a copy of the Valladares contract.

June 18
Terms of the Aranguren concession are amended to include right of petroleum exploitation. Five years later, the Bolívar Coastal field is to be discovered under this lease.

June 30
During the second semester of 1911 and the first semester of 1912, exports of natural asphalt are 15 022 kg, valued at Bs970.

November 27
R. Arnold submits his preliminary geological report to Caribbean Petroleum, recommending the selection of 87 lots of 500 ha each in Monagas, Anzoátegui, Sucre, Nueva Esparta, Falcón, Zulia and Trujillo States. He recommended the immediate drilling of a well in Mene Grande: Zumaque.

December
As a last resort, General Asphalt offers Shell participation in its Venezuelan holdings. Henri Deterding, in 'perhaps the most speculative venture of my life', buys a 51% interest in Caribbean Petroleum for US$10 million.

1913

José María Rivas, in *Trade in Maracaibo* (*El Comercio en Maracaibo*), says Zulia State has three natural asphalt mines (one of 1800 ha) and one of crude oil.

January 17
In a complicated financial deal, Burlington Investment Co. is established, permitting Shell, through Caribbean Petroleum, to acquire control of all petroleum rights in the concessions of General Asphalt (ex-Valladares).

February 1
Caribbean Petroleum starts geological work.

March to June
New York and Bermúdez drills 28 shallow wells in and around the Guanoco asphalt lake, to gain control of the 14 000-ha area.

August 15
New York and Bermúdez discovers the Guanoco field, on completion of well Bababui-1 (*Maturín basin, 60 km northeast of Maturín, first oilfield found in the twentieth century*).

September 3
Venezuelan Oil Concessions obtains the Vigas concession.

October
The Aranguren concession is transferred to Venezuelan Oil Concessions.

November 3
Carlos Alberto Villavicencio asks for a concession on natural gasoline extracted by the distillation of coal.

November 20
B.L. Riera asks for a monopoly of petroleum refining in the principal cities.

1914

January 4
As the exploration period of ex-Valladares second concession expires, Caribbean Petroleum, upon the recommendation of R. Arnold, claims 1028 exploitation lots, covering 512 000 ha around and near oil seeps.

January 8
Caribbean Petroleum contends the Riera monopoly.

March 19
Standard (New Jersey) requests the State Department intervene in relation to the Riera monopoly.

March 30
The Minister of the US Legation, Preston McGoodwin, informs the State Department that the Minister of Development assured him no refining monopoly will be granted to B.L. Riera.

April 15
Caribbean Petroleum discovers the first oilfield of importance in Venezuela, Mene Grande, upon successful completion of new field

Fig. 6. Well Zumaque No. 1 discovered the first giant oilfield in Venezuela, 15 April 1914.

wildcat Zumaque-1 (renamed MG-1). (*Maracaibo basin, 120 km southeast of Maracaibo*).

April 25
The Minister of the US Legation, P. McGoodwin, informs the Secretary of State of an important discovery in Mene Grande.

June 28
First World War. Archduke Franz Ferdinand and his wife are assassinated in Sarajevo.

August 1
Regulations to Mining Law (*Decreto Reglamentario de Minas*) are issued.

August 2
Caribbean Petroleum discovers the small Totumo field on completion of new field wildcat Zambapalo-3 (*Maracaibo basin, 85 km west of Maracaibo*).

1915

February
Shell acquires management control of Venezuelan Oil Concessions.

March 24
The US Consul in Curaçao informs the State Department of machinations to gain control of oil production in Venezuela by the Royal Shell consortium of London and The Hague.

June 26
A new Mining Law (*Ley de Minas*) is enacted.

July 28
The Planas concession of Buchivacoa District is transferred to the Venezuelan Falcón Oil Syndicate.

July 29
Andrés Valbuena and others start long legal proceedings against Caribbean Petroleum.

September 23
Orden-1 is successfully completed by Colón Development as the discovery well of the small Río de Oro field (*Maracaibo basin, 225 km southwest of Maracaibo, just north of the Colombian border*).

1916

March 8
Regulations to the 1915 Mining Law are issued.

July 25
R. Arnold ends his work in Venezuela.

August 27
Colón Development discovers the Las Cruces giant oilfield, on completion of new field wildcat Toldo-1 renamed T-1 (*Maracaibo basin, Tarra group of fields, 255 km southwest of Maracaibo, near Colombian border*).

October 16
The Manager of Caribbean Petroleum reports that the Minister of Development asked for his views on boundary differences with Colombia in the Río de Oro area, his permission to appoint a technical assistant and which articles in the Mining Law are not agreeable to oil companies.

1917

January
Two parallel 16-km long, 20-cm pipelines from the Mene Grande field to San Lorenzo are put into operation.

June 26
The Shops and Public Establishments Law is passed. This is the first move towards regulating working conditions and workers' rights in Venezuela.

August 17
First refining operations in San Lorenzo refinery.

September
First exports of Venezuelan oil by Caribbean Petroleum from the San Lorenzo terminal.

October
Development Minister Gumersindo Torres announces the end of 'blind' concessions for oil exploration.

December 13
Venezuelan Oil Concessions (VOC) successfully completes wildcat Santa Bárbara-1 (renamed R-2). With this completion in the La Rosa area, VOC had unknowingly discovered the Bolívar Coastal field, the third largest petroleum accumulation in the world. Resources of the

field are estimated at 4600 million m^3 of oil (*Maracaibo basin; the La Rosa area is located 40 km southeast of Maracaibo, near Cabimas, on the eastern shore of the lake*).

1918

January 5
Pedro Emilio Coll writes from Madrid to President Gómez on the use of alcohol mixtures for motor gasolines.

June 27
A new Mining Law (*Regimen de la Ley de Minas*) is enacted. It declares that the granting of a concession does not confer ownership of the deposit, but only the right to exploit the substance. Four articles deal exclusively with petroleum, incorporating the ideas of the Development Minister, Gumersindo Torres. The duration of titles was set at 30 years, surface rentals varied in accordance with producing rates, and royalties were fixed at between 8 and 15% depending on the distance between the reservoirs and shipping terminals. Areas not exploited had to be returned to the Government within 3 years. First reference made to conservation measures.

September 15
La Alquitrana field produces 600 gal/month.

October 9
A regulatory decree of the 1918 Mining Law (*Decreto reglamentario del carbón, petróleo y sustancias similares*) is passed. One provision establishes that concessionaires can only exploit half of the area granted for exploration, the other half having to be returned to the State as 'national reserve'.

November 21
The Minister of the US Legation, P. McGoodwin, states that the 'courts in Venezuela are absolutely corruptible'.

December 31
Petroleum appears for the first time in Venezuelan export statistics: 21 194 metric tons, valued at Bs900 000.

1919

Intense competition for concessions develops as Americans are encouraged by the Wilson administration to seek oil rights. Diplomatic support is given to the oilmen by the Caracas US Legation.

December 29
Regulatory decree of the Mining Law.

1920

January 5
The old Planas concession is now transferred to British Controlled Oilfields (Anglo-Persian).

March 17
A second regulatory decree of the 1918 Mining Law is passed.

March 30
Creole Petroleum, a subsidiary of Standard Oil (New Jersey), is incorporated in Delaware.

April 5
The Minister of the US Legation, P. McGoodwin, informs the State Department that President Gómez assured him that the proposal by the Finance Minister, Román Cárdenas, to reduce the size of concessions significantly, will not be accepted by the Government.

June 19
The first Hydrocarbons Law (*Ley sobre Hidrocarburos y demás minerales combustibles*) of Development Minister Gumersindo Torres, is sanctioned by National Congress. Surface taxes are increased, private landholders get an opportunity to obtain exploration permits on their land, the size of concessions is reduced, national reserves are expanded and can only be granted for exploitation, and the list of duty-free import items is greatly reduced. Article 50 fixes the time for selection of exploitation parcels at 3 years.

August 11
International Petroleum, an affiliate of Standard Oil (New Jersey), is incorporated in Canada.

December 31
From 1878, there are 1312 concession contracts for oil exploration and development, of which 835 are in Zulia State.

The production of seven Caribbean Petroleum wells is 69 539 tonnes for the year; asphalt extraction is only 23 452 tonnes.

1921

Alfredo Jahn publishes a summary of the geological formations in Venezuela (*Esbozo de las formaciones geológicas de Venezuela*).

February 15
Settlement of the important 'Vigas concession' case, which might have had far reaching importance concerning the validity of all 1907 leases.

May 2
Pedro Vicente Navarro asks for exploration permits in Cubagua Island.

The US State Department scolds Legation Minister P. McGoodwin for actively helping Colon Development, 75% of which is owned by British subjects.

May 3
Caribbean Petroleum obtains compromise on the controversy over the Aranguren concession. Later, for Bs10 million, any legal arguments on the Valladares leases are settled and Shell has at the end all its holdings guaranteed until 1965 at least.

June 1
More than 2300 Venezuelans have applied for concessions under Article 8 of the Hydrocarbons Law giving landowners an option for leases on their own property.

June 2
A second Hydrocarbons Law (*Ley sobre Hidrocarburos y demás minerales combustibles*) is enacted. President Gómez, yielding to pressure, had allowed the operating companies to help draw up the Law (Minister Torres had been dismissed earlier). Exploitation area was doubled, royalty was fixed at 10%, in cash or kind. Damage, waste and fires were to be avoided.

June 8
The first oilfield in the Falcón basin, El Mene de Mauroa, is discovered by British Controlled Oilfields, on completion of new field wildcat EM-1 (*60 km east of Maracaibo*).

December 5
Standard Oil of Venezuela, a subsidiary of Standard Oil (New Jersey), is incorporated. Former US Legation Minister, P.M. McGoodwin, is now VP-Manager of Creole Oil Syndicate.

1922

January 22
An exploration concession in Cubagua Island is granted to José María Merchán.

March
Two shallow-draft lake tankers, able to clear the Maracaibo bar, are put into operation by Caribbean Petroleum.

April 6
Venezuelan Petroleum (Sinclair) is incorporated in Delaware.

June 9
A third *Ley sobre Hidrocarburos y demás minerales combustibles* is sanctioned by the National Congress. It was a simple, practical document, meant to remove all ambiguities from the 1921 Law. The size of parcels was increased and the duration of exploitation extended to 40 years. Royalties were to be paid on the basis of the market value at embarkation point. All American leases were converted to this liberal Law, which was to regulate the industry for more than 20 years.

December 14
Well Los Barrosos-2 of the Venezuelan Oil Concessions blows out from 500 m, at an estimated rate of 16 000 m^3/day (*La Rosa area, Bolívar Coastal field*). A witness, H. Pittier, estimated the gushing jet to be 30 cm in diameter and reported it could be seen from Maracaibo (?).

Fig. 7. Blow-out of well Los Barrosos No. 2, Sunday 14 December 1922, started rush into Venezuelan oil. After 9 days and 150 000 tonnes, the well sealed itself up. La Rosa area, Bolívar Coastal field.

December 23
Los Barrosos-2 sands up and stops flowing. Total production during the 9-day blow out is estimated at 150 000 m^3.

1923

January 13
A long article in *Deutschezeitung* (Berlin) analyses possible German investments in the Venezuelan oil industry.

February 15
Las Flores-1 (renamed P-3) discovers the La Paz field. The well was completed in shallow Eocene sands (*Maracaibo basin, 40 km west of Maracaibo*).

March 23
Venezuela Gulf Oil is incorporated in Delaware.

June 19
Lago Petroleum starts operations in Venezuela.

June 22
The Compañía Venezolana del Petróleo is incorporated in Caracas.

July 3
The Minister of the US Legation, Willis C. Cook, informs the State Department that the Compañía Venezolana del Petróleo was established as the favoured instrument of Dictator Gómez to dispose of national reserves.

October
Side-stepping Development Minister Antonio Alamo, the representative of Standard Oil in Venezuela receives assurances from President Gómez of a prompt and satisfactory solution to the controversy on the Planas concession.

November
First exports of oil from the El Mene de Mauroa field. Oil was moved to a terminal at La Estacada through a 55-km pipeline laid parallel to the company's narrow-gauge railroad.

December 13
The Federal Executive gives a 25% exemption to exploitation tax payments of Mene Grande.

1924

First geophysical work in Venezuela. It was a short survey by electrical methods, in western Venezuela.

February 5
A German delegation arrives in Venezuela to discuss the Compañía Venezolana del Petróleo with President Gómez.

February 10
President Gómez announces that the German interests, Stinnes, are willing to buy huge holdings of Compañía Venezolana del Petróleo.

February 23
The representative in Venezuela of the German group Stinnes, Wilhelm Waltzing, announces that a 40-day option is secured over 200 000 ha in Maracaibo basin, for US$3 million, with the obligation to complete three exploration wells.

March 26
The US Legation in Caracas is instructed to inform the Venezuelan Government of the concern of the United States if American oilmen are excluded, or discriminated against, on the acquisition of national reserves.

April 22
First well of British Equatorial under waters of Lake Maracaibo.

May
Due to financial difficulties in the Weimar Republic, Hugo Stinnes renounces his options on oil exploration in western Venezuela.

Led by Standard Oil (New Jersey), American companies start buying concessions from the Compañía Venezolana del Petróleo.

June
First oil exports by an American oil company from Venezuela (Lago Petroleum).

June 14
The Government announces prompt offer of national reserves by Compañía Venezolana del Petróleo.

July 24
The Minister of the US Legation, Frederick C. Chabot, informs in detail of a scheme related to oil concessions and notes that most leases are in the hands of US interests.

August 12
Lago Oil and Transport, a wholly-owned subsidiary of Standard Oil (New Jersey), is incorporated in Canada. It owns and operates the Aruba refinery.

1925

First survey in the world with torsion balance completed in Venezuela.

February
A. Hamilton Garner publishes quite a complete picture of the oil geology of northern Venezuela.

March
Lago Petroleum completes a small refinery at La Salina (*near Cabimas*).

May 9
Venezuelan Atlantic Refining is incorporated in Delaware.

June 1
Well R-28 of Venezuelan Oil Concessions blows out and causes furious blaze. Production during the 2-day uncontrolled flow is estimated at 16 000 m^3 of oil (*La Rosa area, Bolívar Coastal field*).

June 2
Caribbean Petroleum discovers a small field at La Concepción on completion of shallow well C-2 (*Maracaibo basin, 20 km west of Maracaibo*).

June 3
Fishermen in La Cañada, Urdaneta District of Zulia State, complain of the pollution of Maracaibo Lake by crude oil from underwater operations.

June 24
The Constitution declares that mines are the property of the States, but are administered by the Federal Government.

July 18
A Hydrocarbons Law, almost identical to that of 1922, is published in the Official Gazette.

July 21
Although not yet properly organized, oil workers strike in the Lake Maracaibo area, protesting against the rapidly increasing cost of living. President Gómez sends troops to maintain order. A 20% gain on the fixed one-dollar-a-day wage is nevertheless gained.

1926

A.H. Garner publishes the first correlation chart of the geological formations in Venezuela.

In a very long communication to the Officio Geologico d'Italia, E. Fossa Mancini reports on an oil seep near Puerto Cabello (apparently, the same one to which Cortese had referred in his 1901 paper).

First seismic surveys in western Venezuela.

Fig. 8. La Paz field, shortly after discovery in 1923.

Subsidence of eastern shore of Lake Maracaibo reported, due to development of Lagunillas area (*Bolívar Coastal field*).

January 18
The Ambrosio area of the Bolívar Coastal field opens up on completion of well Rodríguez-2 of Venezuela Gulf.

March 11
Well Lago-1 of Mene Grande is completed as an oil producer, opening up the Lagunillas area of the Bolívar Coastal field. Although on the basis of an arbitrary nomenclature, Lagunillas is the largest producing area of the BCF.

April
Lago Petroleum acquires the offshore Lake Maracaibo concessions of British Equatorial Oil.

April 13
Very complete report on the oilfields of the Maracaibo basin is presented by C.M. Hunter to the Institution of Petroleum Technologists in London.

May 1
The Venezuelan Ambassador in London, Diógenes Escalante, informs President Gómez that oil wealth is escaping, because of the lack of control on operations of foreign companies.

August 26
Well Unity-1 of Venezuela Gulf opens up the Punta Benítez area of the Bolívar Coastal field.

Texaco Inc. incorporated in Delaware.

October 29
First oil well of Lago Petroleum (LL-2) is completed in the Lagunillas area of the Bolívar Coastal field.

December 31
Petroleum becomes the first export commodity of Venezuela: 3 836 475 metric tons worth Bs192 million.

1927

First geophysical survey in Eastern Venezuela: torsion balance, refraction and magnetic.

January 26
Oil companies are ordered not to drill wells closer than 75 m from each other. This was meant to alleviate wasteful competitive drilling along the Mene Grande strip in Lake Maracaibo, Bolívar Coastal field.

March 4
The Los Manueles oilfield is discovered by Colón Development on successful completion of new field wildcat TM-1, renamed CM-1 (*Maracaibo basin, Tarra group of fields, 15 km north of Las Cruces field*).

April
Mene Grande, Caribbean and Standard (Indiana) agree to reduce competitive drilling and discontinue offsetting.

December 2
A Presidential committee is set up to study the oil pollution of Maracaibo Lake.

December 6
L-44 of Lago Petroleum blows out with tremendous force. Derrick is thrown out into lake.

1928

First book on the geology of Venezuela is published by Ralph Alexander Liddle.

March 9
Lago Petroleum opens up the important Tía Juana area of the Bolívar Coastal field, on completion of wildcat TJ-1.

June
The increasing volume of Venezuelan oil imports begin to disturb US independent producers. Lobbying commences in Washington for a tariff on imports.

June 1
New field wildcat Moneb-1 of Standard Oil of Venezuela, is successfully completed discovering the Quiriquire field. This is a very important discovery, the first in the Maturín basin of eastern Venezuela (*25 km north of Maturín*).

June 7
Standard Oil (New Jersey) purchases its first production of Venezuelan oil, by acquiring control of the Creole Syndicate.

June 15
The oil town of Lagunillas de Agua, on the eastern shore of Lake Maracaibo, is partially destroyed by an oil blaze.

July 11
Congress sanctions a law to prevent the pollution of waters by petroleum.

July 18
A new Hydrocarbons Law (*Ley sobre Hidrocarburos y demás minerales combustibles*) is sanctioned by the National Congress.

July 23
First Labour Law enacted: no minimum wage, unions are allowed, and a maximum working day of 9 hours. Workers can be fined for spreading Communist propaganda.

November 26
Partners in Petrolia give exclusive rights to Clarence John Brown, for final development of La Alquitrana field, use of refinery and sale of products.

December 31
Production for the year surpasses 15 million m^3.

1929

Lago Petroleum uses wooden and concrete foundations in their Lake Maracaibo wells at depths of 15 m.

Dyke built to project eastern shore of Lake Maracaibo against subsidence, due to development of offshore areas of Bolívar Coastal field.

Colón Development completes small refinery at Casigua.

West India Oil – later absorbed by Lago Petroleum – erects small refinery at La Arreaga, Maracaibo.

Petrolia resumes drilling in its La Alquitrana concession.

March
Caracas Petroleum is incorporated in Delaware by Alfred Meyer and others.

March 6
First electric well log outside of France is run in VOC's development well R-216 (*La Rosa area, Bolívar Coastal field*). It was a resistivity survey.

October 29
Stock market crash in the United States.

October 30
Orinoco Oil (Pure Oil) discovers the small Netick field (*Maracaibo basin, 10 km north of La Paz field*). Only field in Venezuela named after an oil man.

November 19
Recently reappointed Development Minister, Gumersindo Torres, increases tax schedule on national reserves by 50%.

December
Oil companies pay Bs12.5 million in back taxes on buoys in the Maracaibo Bar.

December 31
There are 108 oil companies operating in the country, including Venezuela Speculation Inc., but only five export significant volumes of oil.

1930

July 16
The Ministry of Development establishes the Technical Office of Hydrocarbons (*Oficina Técnica de Hidrocarburos*).

August 7
Regulations (*Regulamento sobre la Ley*. . .) for the 1928 Hydrocarbons Law are issued by Development Minister Torres, to protect in an efficient manner the interests of the Nation in the petroleum industry.

September 20
In a memorandum to the Ministry of Development, oil companies vehemently protest governmental interference as laid out in the new Regulations.

October 30
Development Minister Torres informs oil companies that attacks on the Regulations would be tantamount to denying the sovereignty of the country. Certain impractical provisions are suspended, but all new procedures on fiscalization and supervision by Government field inspectors are maintained.

November
As activities are drastically reduced, and thousands dismissed, oil workers of Mene Grande try unsuccessfully to strike.

November 24
The Government sends a group of Civil Engineers, graduates of the Caracas Central University, to the United States to specialise in oil.

December
Shell announces a voluntary decrease in production of around 20%; Mene Grande and Standard (Indiana) 10%.

December 7
Blow out of wildcat Lagunita-1, of Mene Grande, the first well of the Bachaquero area of the Bolívar Coastal field.

Minister Torres announces change of policies with reference to the oil industry.

1931

The SP curve of electric logs introduced commercially in Venezuela.

March
In order to ease the situation around the oilfields, the National Government offers to transport back to their homes those dismissed by the operating companies.

March 22
A group of exiled Venezuelans publish an intended programme of government. Barranquilla (Colombia) Plan, Item 7, asks for a thorough revision of the concessions policy.

March 31
Development Minister Torres appoints the geologist Guillermo Zuloaga, General Inspector of Hydrocarbons in the Technical Office of Hydrocarbons.

April
Mene Grande, Shell and Standard (Indiana) agree to restrict production by 15%.

April 24
The Cumarebo oilfield is discovered by Standard Oil of Venezuela (Creole). Although small it is the largest field found in the Falcón basin (*30 km E of Coro*).

June
Development Minister Torres accuses the oil companies of misrepresenting financial information to the Government.

August
No well completions during the month – an effect of the depression.

August 22
C.J. Brown reports that there is no hope for further developments in La Alquitrana field.

September
Pres. Gómez, yielding to pressure, dismisses Minister Torres. Claims for payments due are not pressed.

October 13
Standard Oil of Venezuela inaugurates the Caripito refinery, which would be for many years the largest in the country.

1932

May
Standard Oil (New Jersey) buys lake concessions of Standard Oil (Indiana), thus becoming the dominant oil concern in Venezuela.

June 1
Cumulative production in Venezuela tops 100 million m^3.

June 6
The US Congress passes a law introducing a crude and fuel oil tariff of US$0.21 per barrel and a gasoline tariff of US$1.05, for all imports into the United States.

November 18
The first electric log in eastern Venezuela is run in well Q-78 Quiriquire field.

December 31
There are more than 1000 oil-producing wells in the country, for the first time.

1933

Gas injection starts in the Cumarebo field.

Pressure maintenance by gas injection starts in the La Rosa area of the Bolívar Coastal field.

April
Lago Petroleum starts using concrete pile foundations for offshore drilling in Lake Maracaibo. Piles are made in a plant in Lagunillas.

Creole

Fig. 9. End of La Alquitrana field of Petrolia.

April 15
Standard Oil of Venezuela discovers the Orocual field on completion of well Orocual-2 (*Maturín basin, 20 km southwest of Quiriquire*).

June 20
The Pedernales giant oilfield is discovered by Standard Oil of Venezuela (Creole) on completion of new field wildcat Pedernales-2 (*Maturín basin, 100 km north-northeast of Maturín*).

September 2
The US Secretary of Interior announces imports of crude petroleum

are to be limited to an amount not exceeding the average daily imports during the second half of 1932. Voluntary compliance to the programme is requested.

November 3
A 'B/L' agreement is signed by Mene Grande and Standard Oil of Venezuela. No details available.

1934

April 8
Petrolia ceases operation, as 1878 concession lapses.

June
Lago Petroleum plants in La Salina start producing steel and concrete piles for well foundations offshore in Lake Maracaibo.

July 25
President Gómez sets up subsidy for agricultural exporters, as bolívar appreciates to a rate of Bs3 to US$1.

August 23
Exchange controls are introduced. By agreement with the Government, oil companies would sell two-thirds of their dollars to Venezuelan banks at Bs3.90, one-third to the Government at Bs3.06.

December 12
Proved reserves stand at 400 million m^3.

1935

May 27
Import quotas introduced in 1933 in the United States end with death of the National Industry Recovery Act. Voluntary restraint on imports continues.

May 30
Ultramar Exploration is formed in London by Alfred Meyer and associates.

June 17
The National Congress enacts a new Hydrocarbons Law (*Ley sobre Hidrocarburos y demás minerales combustibles*).

July
Socony tries to bid on national reserve lots offered by the Government, but Minister of Development, Cayama Martínez, gives them to Standard Oil (New Jersey) companies.

November 4
Regulations (*Reglamento sobre la Ley . . .*) for the 1935 Hydrocarbons Law are issued.

December 17
President Gómez dies and General Eleazar López-Contreras temporarily assumes executive powers.

1936

Subsidence reported onshore the Bachaquero area (*Bolívar Coastal field*).

January 7
First well to drill through the 'tar belt' of the southern border of the Maturín basin, La Canoa-1, is completed (*145 km southwest of Maturín, 50 km north of the Orinoco river*).

February 27
Establishment of the first oil workers' union, *Sindicato de Obreros y Empleados Petroleros de Cabimas.*

March 20
Establishment of the Consultative Department for Mining and Geology, ascribed to the Ministry of Development. S.E. Aguerrevere is appointed first Director of the Department.

April 25
General López-Contreras is elected President of the Republic by the National Congress.

May
Development Minister, Néstor Luis Pérez, announces that concessions will be granted on competitive bidding. Companies raise their royalty payments to 15%, offer rental and surface payments at 10 and 15

times the minima, agree to pay in bolívars purchased at Bs3.06 (not the official fixed rate of Bs3.90), and to process larger percentages of their output in Venezuela.

June 15–23
Public debate on oil policies in Caracas daily *La Esfera* between Rómulo Betancourt and Alejandro Pietri, lawyer of Standard Oil of Venezuela.

July 14
Arturo Uslar Pietri, in an editorial in the Caracas daily *Ahora*, puts forward the idea that Venezuela needs 'to sow the petroleum' – *sembrar el petróleo*.

July 16
The Labour Law is sanctioned. It includes the most advanced principles of modern social law.

July 20
A law is passed concerning pollution of waters by oil.

August 5
A new Hydrocarbons Law (*Ley sobre Hidrocarburos y demás minerales combustibles*) is sanctioned by the National Congress. Special advantages of a non-monetary nature could be asked for from companies willing to receive concessions.

September 14
Sharp attacks in the newspaper *Crítica* on the 'corrupt methods' of Standard Oil (New Jersey) in acquiring Venezuelan concessions and also on the legality of the Valladares lease.

September 15
New field wildcat Temblador-1 (TT-1) of Standard Oil of Venezuela discovers the Temblador giant oilfield (*Maturín basin, 100 km south of Maturín*).

October
Minister Pérez demands that oil companies clean up the oil contaminated shores of Lake Maracaibo and that 'company roads' be open to general circulation.

November 7
Wildcat Santa Ana-1 (renamed AM-1) of Mene Grande makes the first discovery of the Anaco Group of fields (*Maturín basin, 50 km southwest of Maturín*).

December 14
Thriving on the privileges obtained through the new Labour Law, recently established unions call for a general strike.

1937

Following the discovery and rapid development of Oficina and fields in the central area of Anzoátegui State, El Tigre evolves into an active population centre. In twenty-five years it would be the second largest town in Eastern Venezuela.

The Uracoa field is discovered.

First issue of *Boletín de Geología y Minería.*

January 15
Regulations on the 1936 Hydrocarbons Law are issued (*Reglamento sobre la ley . . .*)

January 18
IBM starts operating.

January 22
President López-Contreras decrees an end to the oil workers' strike. Operating companies do not recognize unions nor give basic rights for payment of overtime and vacations to the workers. Government grants a Bs1 increase in wages.

January 27
Development Minister N.L. Pérez again raises a claim for incorrect reporting of transport costs. Standard settles its Bs26 million claim for a little more than Bs4 million.

February 15
The First Venezuelan Geological Congress opens in Caracas.

The Technical Service of Mining and Geology publishes the first preliminary geological map of northern Venezuela.

Creole

Fig. 10. Early development drilling in Quiriquire field, eastern Venezuela.

April 14
MG-1 (Merey-1), of Mene Grande, discovers an important oilfield (*Maturín basin, 35 km southeast of Oficina*).

April 22
Standard Oil of Venezuela discovers the Pilón oilfield (*Maturín basin, 10 km south of Temblador*).

June 16
Oficina-1, the discovery well of the Oficina field, renamed OG-1, is completed for 212 m^3/day of oil (*Maturín basin, 150 km southwest of Maturín*).

July 27
An investigating commission of the National Congress condemns the practice of companies informing the Government of the names of the 'agitators', causing 30 labour leaders to be expelled from the Lake Maracaibo area.

September 29
The Geological Institute is created by Presidential Decree.

November 2
New field wildcat Yopales-1 (renamed YS-1) of Mene Grande is successfully completed, discovering a giant oilfield (*Maturín basin, Oficina group 20 km southwest of the Oficina field*).

December 15
Mene Grande and Standard Oil of Venezuela sign an agreement in Caracas which in effect transforms the wholly-owned Gulf subsidiary into a joint enterprise.

An agreement is signed in Toronto between Mene Grande and International Petroleum, a subsidiary of Standard Oil (New Jersey). It was known among the parties as the 'principal agreement'; it was announced to the public as a contract of Standard to buy oil from Mene Grande. The principal agreement gave International supervisory control on all Mene Grande activities, including a veto over major projects. International acquired half of the physical properties of Mene Grande and half of all future production from Mene Grande concessions in Venezuela for US$125 million.

Mene Grande and the two subsidiaries of Standard Oil (New Jersey) operating in Venezuela, Lago Petroleum and Standard Oil of Venezuela, sign an agreement in Toronto fixing production quotas for all three companies for the next 12 years. International also participates in the agreement by virtue of the 'principal agreement'. The 'ratio agreement' provided for an output of Standard–Lago 3454 times that of Mene Grande.

1938

January
President López Contreras decides no more oil concessions should be given so that 'full benefits rightful to the nation could be received in

the future'; amid agitated public debate over oil issues and the role of foreign companies in the exploitation of Venezuelan oil, the Government announces suspension of the granting of concessions.

March 18
President Lázaro Cárdenas of México signs an Executive Decree expropriating foreign oil properties.

March 20
Exploratory well Tigre-2 (LG-1) of Mene Grande discovers the giant Leona oilfield (*Maturín basin, Oficina group, 55 km northeast of the Oficina field*).

April 4
The Supreme Court orders Mene Grande to pay Bs16 million for collecting illegal rebates on some concessions granted during the mid-1920s.

April 6
Second Geological Congress ends in San Cristóbal.

July 13
After 2 months of violent debate, the National Congress passes a Hydrocarbons Law (*Ley sobre Hidrocarburos . . .*) containing notable differences to previous laws. The State is authorised to carry on any oil activities. The Law provided for royalties of up to 16% in national reserves, a 40-fold increase of the exploration tax, and a 4-fold increase of the initial exploitation tax. Natural gas should be conserved, and a new method of fixing the market value of the oil should be formulated.

September 16
Inauguration of the Geological Institute (Ministry of Education), now the Geology, Mining and Geophysics School of the Central University of Caracas.

October 13
Standard Oil of Venezuela completes new field wildcat Jusepín-1, discovering one of the largest fields in Venezuela (*Maturín basin, 35 km west of Maturín*).

October 22
Caracas Petroleum, a subsidiary of Ultramar, is incorporated in Caracas.

November 30
Regulations of the 1936 Labour Law are issued.

International and Nederlandsche Olie Maatschappij, a subsidiary of Shell, sign an agreement in Hudson County, NJ, whereby International sold half of the interests it had acquired the previous year from Mene Grande. The 'main agreement', retroactive to 15 December 1937, in fact made NOM(Shell) a full and equal partner with International in the previous arrangements.

International and Shell (through NOM) sign a second document in which Shell recognises, and assents to, the ratio agreement.

December 19
Mene Grande 'reserves its rights' referring to the main agreement between International and NOM.

December 21
The National Congress enacts a new Hydrocarbons Law (*Ley sobre Hidrocarburos y demás minerales combustibles*).

1939

Shell uses palinology (pollen and spores) for the first time in oil exploration. A small laboratory is set up in Maracaibo.

Subsidence first reported in Tía Juana (*Bolívar Coastal field*).

January 28
In two separate documents, Mene Grande recognises the right of International to admit NOM(Shell) to the arrangement signed concerning the production and sharing of Venezuelan crude.

April 4
Creole discovers the giant San Joaquín oilfield on completion of new field wildcat JMN-1 (*Maturín basin, Anaco group, 150 km southeast of Maturín*).

April 5
The Supreme Court rules that Mene Grande should pay Bs15.5 million, 25% of the exploitation tax exempted in 1923.

May 10
Wildcat RPN-1 of Pantepec (Phillips) discovers the El Roble field (*Maturín basin, Anaco group, 12 km northeast of San Joaquín field*).

May 29
A commission of the Ministry of Development studies the most convenient site for a national refinery near Puerto La Cruz.

June 13
Mene Grande completes Oficina well OG-24, the first dual completion in eastern Venezuela.

September
A 24-cm pipeline is put into service between the Jusepín field and Caripito by Standard Oil of Venezuela.

September 1
Second World War. Adolf Hitler, Führer of Germany, declares war on Poland.

September 8
Venezuelan Central Bank established to control foreign exchange.

October 7
Creole discovers prolific Eocene production in the Bolívar Coastal field on completion of new pool wildcat LL-370, offshore in Lake Maracaibo.

October 22
The First Petroleum Industry Exhibition is inaugurated by President López-Contreras.

November 6
Venezuela and the USA sign a commercial treaty, the Reciprocal Trade Agreement. Tariff on petroleum exports is reduced by half. Venezuela receives approximately 90% of the import quota.

December
First shipment of crude through 33-cm 155-km long pipeline from Oficina to Puerto La Cruz. At the time, it was the longest pipeline of its size in the world.

December 18
President López-Contreras opens First Conference of Government Petroleum Inspectors, in Caracas.

December 31
Yearly production reaches 30 million m^3 for the first time.
Nearly 2 million m^3 of crude are processed in Venezuelan refineries.

1940

Two dry holes are drilled in the island of Cubagua.

January 1
Exchange rate of oil dollar fixed at Bs3.09 by National Office of Exchange (*Oficina Nacional Centralizadora de Cambios*).

January 24
Regulations to the 1938 Hydrocarbons Law are issued (*Reglamento sobre la Ley . . .*). Development Minister Manuel Egaña declares that the Regulations apply to all concessionaires.

April 2
Oil companies object to Regulations of the 1938 Hydrocarbons Law – in particular, to the need of a government permit in order to drill a location, to begin new construction and to transfer titles of concessions.

April 17
Development Minister Manuel Egaña rejects objections of oil companies to the Regulations of the 1938 Hydrocarbons Law.
Creole completes El Roble-2 (RNP-2), bottomed at 3052 m, then the deepest hole in South America (*El Roble field*).

September 12
Socony makes an important discovery on completion of exploratory

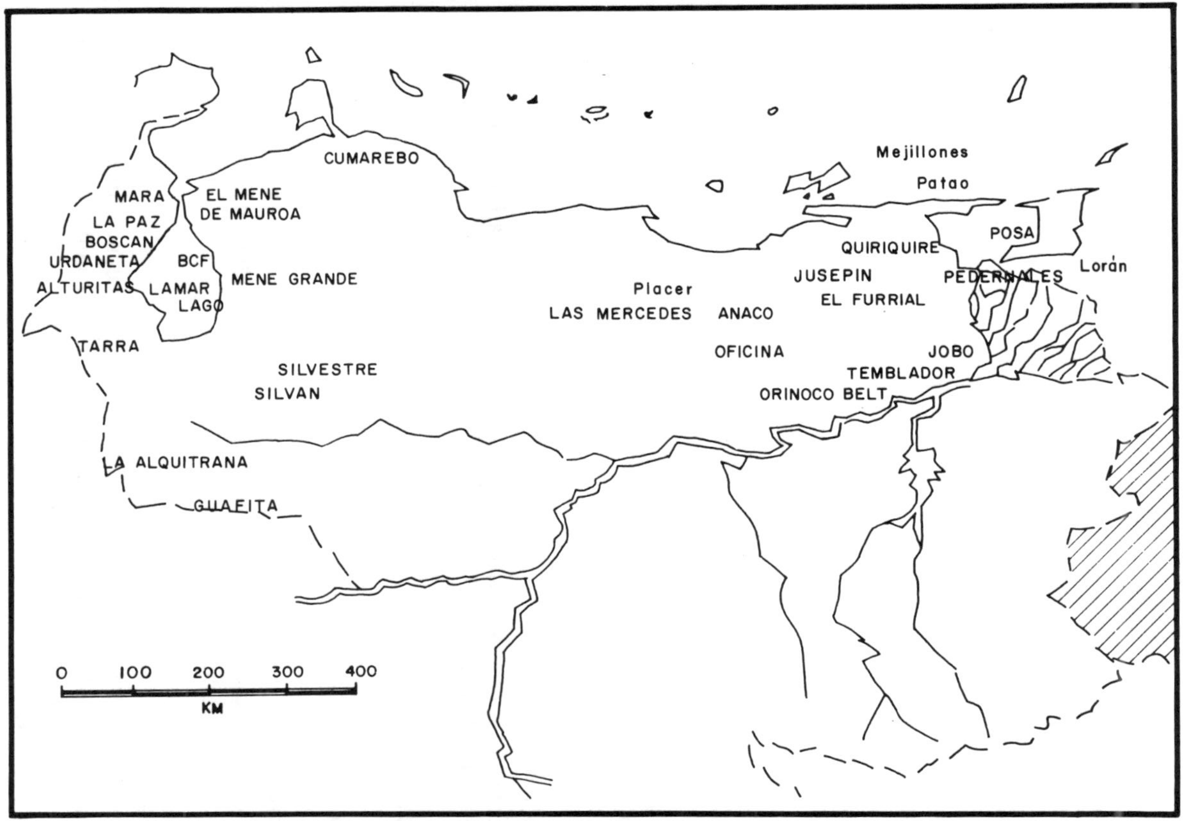

Fig. 11. Oilfields (in capital letters); gasfields (in lower-case letters).

well Guario-1 (*Maturín basin, Anaco area, 15 km northeast of San Joaquín*).

October 15
Authority for fixing oil dollar exchange rate given to Central Bank.

December 30
The North Oficina field is discovered by Mene Grande on completion of wildcat OG-116X (*Maturín basin, Oficina group, 7 km north of the Oficina field*).

1941

Maximum subsidence of 18 cm/year measured in the Tía Juana area (*Bolívar Coastal field*).

January 24
Wildcat RG-2 of Mene Grande discovers the giant Santa Rosa oilfield (*Maturín basin, Anaco group, 25 km northeast of San Joaquín*).

March 7
Consolidada de Petróleo (Sinclair) opens up the Santa Bárbara area of the Jusepín field.

March 13
The Pueblo Viejo area of the Bolívar Coastal field opens up on completion of VOC's well LS-645.

April 28
General Isaías Medina-Angarita is elected President of the Republic by the National Congress.

May 23
On settlement of a claim raised by the Development Minister, Néstor Luis Pérez, in 1937, Mene Grande pays Bs30 million for incorrect reporting of transportation costs.

The Trico giant oilfield is discovered by the Mene Grande on completion of well OG-168 (*Maturín basin, Oficina group, 20 km north of the Oficina field*).

July 23
Exchange rates are fixed at Bs3.09 for oil companies, Bs3.35 for the public, Bs4.30 for cacao and Bs4.60 for coffee (for US$1).

October 7
President Medina sends an economic delegation to the USA to ask for special consideration regarding war export restrictions, since smooth operation of the vital and strategic oil industry of Venezuela was of particular importance to the Allied Forces. The mission was unsuccessful.

November 25
The Las Mercedes giant oilfield is discovered by Las Mercedes on completion of new field wildcat Mercedes-2 (*Maturín basin, western end, 150 km south of Caracas*).

December 26
Creole opens up the Mulata area of the Jusepín field on completion of MC-1 (ex MP-1).

December 31
Daily production rate passes 100 000 m^3.

1942

February 14
Seven tankers in the Lake Maracaibo–Curaçao and Aruba run are torpedoed by German submarines.

February 26
Venezuela and Great Britain sign an international treaty delineating the submarine areas in the Gulf of Paria, between Venezuela and Trinidad. This is the first agreement in the world relating to the continental shelf.

June 2
Mene Grande makes an important discovery on completion of wildcat GG-1 the giant West Guara field (*Maturín basin, Oficina group, 20 km northeast of the Oficina field*).

June 15
President Medina receives extraordinary powers to secure the smooth operation of the oil industry.

July 16
President Medina announces that the petroleum legislation is being revised to ensure the State a greater and more just participation in the subsoil riches of the country.

Minister Pedro Ignacio Aguerrevere signs first agreements with oil companies establishing the commercial value of Venezuelan crudes, in comparison with the actual price of similar crudes in East and West Texas.

July 17
Income taxes are introduced. The first Income Tax Law, to become effective on 1 January 1943, is issued. Oil companies are taxed at a rate of 2.5% of their net income (cedular tax) plus a progressive tax related to the amount of net earnings.

September
First class of Geological Institute graduates.

September 7
An important discovery – the East Guara field – is made by Mene Grande on completion of new field wildcat GG-2 (*Maturín basin, Oficina group, 30 km northeast of the Oficina field*).

October 27
The giant Oveja (previously Aventazón) oilfield is discovered by Mene Grande on successful completion of new field wildcat OM-2X (*Maturín basin, Oficina group, 20 km south of the Oficina field*).

November 15
President Medina reaffirms in Maracaibo the stand of the Executive with respect to the forthcoming petroleum legislation; he said the Government intended to obtain a more equitable share from the exploitation of the oil.

November 20
Consolidada (Sinclair) opens up the Muri area of the Jusepín field.

December 31
Production during the year drops 25% as compared with the previous year. Following shelling of the Aruba and Curaçao refineries by German submarines, a convoy system had been set up. However, tankers were transferred to Europe and the Pacific, while in Venezuela oil companies could not obtain scarce materials to carry out their programmes.

1943

January 13
Prior to the merging of all interests of Standard Oil (New Jersey) in Venezuela, Lago Petroleum, Mene Grande, Standard Oil of Venezuela and Nederlandsche Olie Maatschappij execute an agreement cancelling the ratio agreement of 15 December 1937, as of 31 December 1942.

January 30
The United States and Mexico sign an agreement which in effect allows unlimited volumes of Mexican oil to enter the USA overland.

February 12
President Arthur Proudfit of Standard Oil of Venezuela announces oil companies are in agreement with new petroleum law.

March 5
A congressional commission reports that title to a number of concessions, including that of Valladares and some of Mene Grande in Lake Maracaibo, is contestable.

March 13
National Congress sanctions new Hydrocarbons Law. It was meant to be a mutually profitable arrangement; the Government would receive a greatly enlarged share in the profits (including minimum royalties of $16\frac{2}{3}$%), in exchange for 40 additional years of exploitation, the conversion of all concessions previously granted to the Law – clearing any claims on the validity or legality of many of them – and the promise of the granting of extensive new areas, for the oil companies. The Law promoted domestic refining and broadened the technical and supervisory powers of the Government.

March 16
The General Administration for Income Tax (*Administración General del Impuesto sobre la Renta*) is established.

March 21
Organic decree of the Technical Office for Hydrocarbons (*Oficina Técnica de Hidrocarburos*) is issued.

August 19
All interests of Standard Oil (New Jersey) in Venezuela are consolidated in Creole Petroleum Corporation.

August 31
The regulations to the Hydrocarbons Law are issued.

September 15
Mene Grande well YS-2 discovers the Agua Clara (now, the North Yopales) field (*Maturín basin, Oficina group, 20 km west of the Oficina field*).

October 15
The Ostra field is discovered by Mene Grande on completion of wildcat well OM-3X (*Maturín basin, Oficina group, 15 km southeast of the Oficina field*).

December 31
Following the sanction of the new Hydrocarbons Law, oil companies convert 6 million ha to it and give up 2 million ha.

Proved reserves of the country are estimated at 916 million m^3 of oil. From this date on, proved reserves data are published regularly as of year-end.

1944

March 24
The Workers' National Convention (*Convención Nacional de Trabajadores*) is dissolved by decree.

April 21
President Medina starts gigantic concession-granting programme. In

Fig. 12. Lagunillas busy launch terminal, offices and camp, early 1940s. Looking eastwards.

less than 1 year more acreage is given than that previously held by oil companies. The total area granted was more than 6·5 million ha.

April 28
Richmond Exploration, subsidiary of Standard Oil (California), is incorporated in Delaware.

May 3
Deeper pool test P-62 in the La Paz field is completed as the discovery well of prolific Cretaceous production; so the field is recognised as a giant accumulation.

June 20
Socony makes an important discovery in Eastern Venezuela on completion of wildcat Güico-1 (*Maturín basin, Oficina group, 5 km west of the West Guara field*).

November 10
President Medina decrees a Bs2 increase in pay, thwarting the strike of the reorganised labour unions. No social benefits granted.

December
Mene Grande and Creole complete a 38 cm pipeline from the Jusepín group of fields to Puerto La Cruz.

1945

March 2
Completion of wildcat DM-2 of Caribbean Petroleum, discovering the giant Mara oilfield. The first development areas were designated *Los Tetones* and *Kilómetro 24* (*Maracaibo basin, 35 km northwest of Maracaibo, 20 km northeast of the La Paz field*).

May 14
New field wildcat Texas-1 of Texas discovers the Tucupita oilfield (*Maturín basin, 80 km east of the Temblador field*).

June 14
An agreement is signed in the Labour Inspector's office in Maracaibo between representatives of oil workers and the operating companies, whereby certain minimum conditions are obtained by the unions.

September 26
The concessions-granting cycle which followed the passing of the 1943 Hydrocarbons Law, is closed.

October 18
President Medina is overthrown. Rómulo Betancourt heads Revolutionary Junta.

December 8
A second agreement, leading to the first Collective Contract, was signed in Caracas between oil workers and company representatives, whereby certain additional benefits are obtained by the unions.

December 12
By resolution of the National Commission of Consumer Goods (*Comisión Nacional de Abastecimiento*) maximum prices are set for gasoline and kerosene sold on the domestic market. At Bs0.10 per litre, gasoline prices are the lowest in the world.

December 18
Wildcat NG-1 of Socony discovers the giant Nipa oilfield (*Maturín basin, Oficina group, 10 km north of the West Guara field*).

December 31
The Revolutionary Junta decrees an extraordinary tax of Bs89 million on company earnings.

Production exceeds 50 million m^3.

1946

January
First directional drilling (*Lagunillas area, Bolívar Coastal field*).

Water injection for pressure maintenance is started in the Oficina field.

January 14
Mene Grande discovers the Caico Seco field on completion of well Caico Seco-1, renamed CaM-1X (*Maturín basin, Oficina group, 35 km northwest of the Oficina field*).

January 24
President Betancourt declares that the 1943 Hydrocarbons Law will be respected.

February 26
Sinclair completes first well in the Travieso area of the Jusepín field.

March 7
Socony discovers the West Güico field on completion of wildcat WGV-32 (*Maturín basin, Oficina group, 5 km west of the Güico field*).

March 30
One of the most powerful federations in the Venezuelan Confederation of Labour, *Fedepetrol* (Petroleum Workers Federation), is established at an oil workers' congress in Caracas.

May 17
The South Güico field is discovered by Socony on completion of well

SGV-39 (*Maturín basin, Oficina group, 10 km south of the Güico field*).

June 14
The First Collective Contract is signed. The unions obtain payment for the day of rest and 15 days' paid annual leave.

September 23
Richmond discovers the Ensenada condensate field on completion of wildcat Zulia 1K-1 (*Maracaibo basin, 35 km south of Maracaibo*).

October 2
Phillips discovers the Mata Grande oilfield on completion of well FT-1 (*Maturín basin, Jusepín group, 10 km west of the Santa Bárbara area of Jusepín field*).

October 17
The Boscán field, one of the largest accumulations of oil in Venezuela, is discovered by Richmond on completion of new field wildcat Zulia 7F-1 (*Maracaibo basin, 50 km southwest of Maracaibo*).

December 30
The Income Tax Law is amended, increasing the maximum rate to 26%.

December 31
Gas production is over 10 000 million m^3 during the year.

1947

February 21
Venezuela ratifies the Bretton Woods agreement and becomes a member of the International Monetary Fund.

March 18
Mene Grande completes the first electrostatic dehydration installations at Lagunillas.

April 7
New field wildcat GXB-6 of Creole discovers the Lechozo gasfield (*Maturín basin, 25 km northwest of Las Mercedes*).

June 12
Important Cretaceous production is discovered by Colón Development, the West Tarra field, on completion of wildcat WT-2 (*Maracaibo basin, Tarra group, 15 km southwest of the Las Cruces field*).

June 17
A condensate recovery plant goes on stream in the Cumarebo field.

June 26
Minister Juan Pablo Pérez Alfonzo announces that 25% of the 1948–9 royalties, some 6 million m^3 of oil, would be marketed directly by the Government.

September 17
Texaco Maracaibo domiciled in Venezuela.

Fig. 13. Concessionaires had to complete other operations, such as building of Coromoto hospital in Maracaibo.

October 6
The first of five bids to sell the royalty oil offered by the Government is signed.

December 16
Atlantic discovers the Sabán field on completion of wildcat Sabán 2-1A (*Maturín basin, 65 km east of the Las Mercedes field*).

December 31
The number of oil-producing wells in the country exceed 5000 for the first time. Oil production exceeds 200 000 m^3/day for the first time. Cumulative natural gas production exceeds 100 000 million m^3.

1948

January 17
Socony discovers the giant Chimire field on completion of new field wildcat Chimire-2, renamed CHV-2 (*Maturín basin, Oficina group, 15 km north of the Oficina field*).

January 22
Creole agrees to pay US$0.24 per barrel, instead of US$0.19, for the remainder of its royalty oil.

February 1
Socony makes the first discovery in the Barinas basin on completion of new field wildcat San Silvestre-2. The field is later recognised as a giant (*35 km southeast of Barinas*).

February 13
A 2-year barter deal, involving 325 000 m^3 of Venezuelan oil and 5000 tons of frozen beef from Argentina, is signed.

February 15
Rómulo Gallegos is inaugurated President of the Republic.

February 20
Second Collective contract signed. Wages were increased, commissary prices were frozen at cost, 3 weeks' paid annual leave was granted, and housing and many other social benefits were also increased.

February 22
Mene Grande discovers the El Toco oilfield on completion of new field wildcat Toco-1, renamed TM-1 (*Maturín basin, Anaco group, 20 km west of the Santa Ana field*).

March 11
Former Development Minister, M. Egaña, E. J. Aguerrevere and Deputy Alberto Carnevali are nominated to study the convenience and possibility of building a National Refinery and of exploiting national reserves adjacent to existing fields.

April
The US Secretary of Defense proposes a governmental agreement to use US troops, if necessary to defend the Venezuelan oilfields, which is considered unacceptable by the US Ambassador in Caracas.

April 29
Collective contract for marine officers signed.

May 4
Collective contract with 'marinos' signed.

June
First Cretaceous production in the La Concepción field.

Las Mercedes completes a 38-cm, 250-km pipeline from the Las Mercedes field to Pamatacual.

June 6
The US Ambassador in Caracas proposes to train US personnel in antisabotage operations, to protect the oilfields.

June 16
Richmond discovers the San José field on completion of new field wildcat Zulia 36E-1 (*Maracaibo basin, 110 km southwest of Maracaibo*).

July 1
The Technical Commission of Mines and Geology assumes the name Institute of Mines and Geology (*Instituto de Minería y Geología*).

1948

September 8
The Venezuelan Association for Geology, Mining and Petroleum (*Asociación Venezolana de Geología, Minería y Petróleo*) is founded in Caracas.

September 12
The Ganso field is discovered by Mene Grande on completion of well MFS-1, renamed MS-601X (*Maturín basin, Oficina group, 10km south of East Guara*).

Atlantic discovers the Placer gasfield (*Maturín basin, 50km northeast of the Las Mercedes field*).

September 16
The US Ambassador in Caracas alerts President Gallegos to communist activities in the Puerto La Cruz and Paraguaná refineries.

September 21
Pressure maintenance by gas injection started in Las Mercedes.

October 14
Congress approves report of Permanent Commission for Development referring to Minister's Memoir on the oil industry – specifically, the no-concession policy, the establishment of a national oil company and the possibility of direct sale of royalties.

November 12
National Congress passes a new Income Tax Law, creating the '50–50' sharing of profits. Article 31 establishes an additional tax for the extractive industries of 50%, applicable when oil companies' income, after payment of taxes, exceed the Government's total revenue.

November 24
President Gallegos is ousted; Lt. Colonel Carlos Delgado-Chalbaud heads Military Junta.

December 10
Shell discovers the Sibucara field on completion of well S-5. Portions of the proved area of the field are within the city limits of Maracaibo. At the time it was one of the deepest producing fields in the world.

December 14
Creole completes a 230-km pipeline from Ulé, on the eastern shore of Lake Maracaibo, to Amuay, in the Paraguaná peninsula.

December 31
Area under concessions exceeds 7 million ha.

1949

January 19
A People's Liberation Committee declares a general strike of petroleum workers throughout Monagas State.

January 27
The US Ambassador in Caracas reports on the strategic importance of the Venezuelan oil industry.

February 1
Wildcat B-1 of Phillips discovers the San Roque field (*Maturín basin, Anaco group, 15 km west of the San Joaquín field*).

The Cardón refinery of Shell starts operations. This, and four other refineries in the country, were built to comply with provisions of the 1943 Hydrocarbons Law.

February 9
Decree 40 establishes the National Commission for Mining and Oil Policies (*Comité Nacional de Política Minera y Petrolera*).

Labour policy of the Military Junta is to end political influences in the worker's unions.

February 23
Secretary General of Fedepetrol, Luis Tovar, in a letter to the Military Junta, requests that all union officials imprisoned since November be liberated.

February 25
A general strike for an indefinite period is ordered by the Venezuelan Confederation of Labour. Decree 56 of the Military Junta dissolves the Confederation and orders new elections for all trade unions, associations and leagues belonging to it.

1949

April 7
The Military Junta invites the International Labour Office to investigate the freedom of association and the conditions of work in Venezuela.

April 25
Regulations to the Income Tax Laws are issued.

June
J. E. Pogue, then Vice President of the Chase Manhattan National Bank, refers in a paper to the advantages of oil from the Middle East compared with that of Venezuela. As a solution he suggested that the tax arrangements of 1948 be eliminated, that new concessions be granted to the international majors, that no attempt be made to increase the volume of oil refined in the country, and that no further social or economic benefits be granted to the petroleum workers.

June 9
The Ruiz field is discovered by Atlantic and Pancoastal on completion of wildcat Ruiz 3–1 (*Maturín basin, 45 km southeast of the Las Mercedes field*).

July 10
Mene Grande discovers the important Mapiri oilfield on completion of wildcat well SG-1 (*Maturín basin, Oficina group, 25 km north of the Chimire field*).

July 29
Well SM-200 of Mene Grande is successfully completed as the discovery well of the Soto giant oilfield (*Maturín basin, Oficina group, 20 km north of the Chimire field*).

August 30
A mission of the International Labour Office concludes its visit to Venezuela, after studying working conditions and freedom of association in the country.

September
Minister of Foreign Affairs, L. E. Gómez-Ruiz, and Development Minister, M. Egaña, appoint Edmundo Luongo Cabello, Luis Emilio

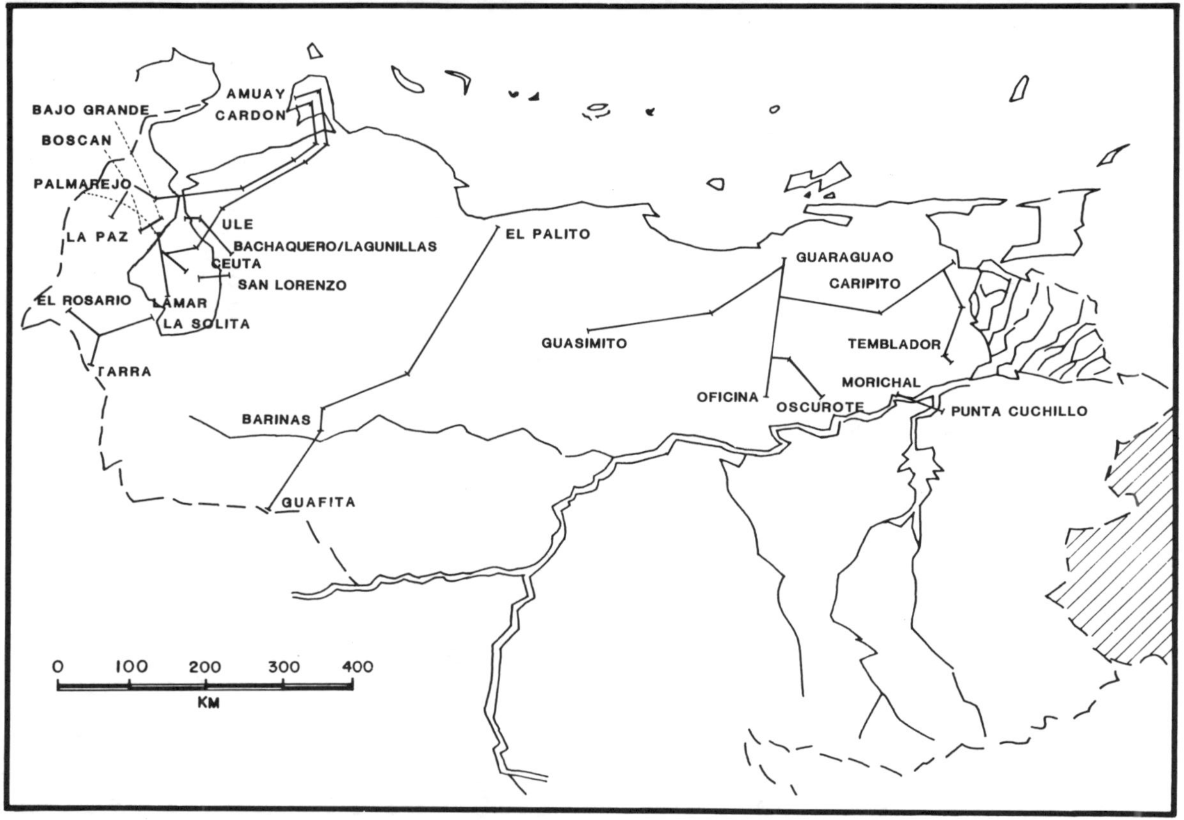

Fig. 14. Oil pipelines.

Monsanto and Ezequiel Monsalve Casado to a Special Mission for the purpose of paying an official visit to Saudi Arabia, Iran, Egypt, Iraq, Kuwait and Syria. This is the first direct contact between Venezuela and the oil-producing countries of the Middle East.

October 1
The agreements concluded by the Ministry of Development and concessionaire companies on the determination of the market value of Venezuelan crude are terminated (except Creole: 31 October 1949).

December 2
The US Federal Trade Commission orders an investigation of the international cartel activities in the oil industry.

December 31
Proved natural gas reserves are 233 473 million m^3. From this date, year-end estimates are produced by the Ministry of Mines and Hydrocarbons.

1950

First tectonic map of Venezuela published.

January 4
The Amuay refinery of Creole goes on stream.

March 7
The Belén gasfield is discovered by Las Mercedes (*Maturín basin, 15 km south of the Las Mercedes field*).

April 9
Atlantic discovers the Oritupano giant oilfield on successful completion of new field wildcat Oritú-1 (*Maturín basin, Oficina group, 40 km east of the Leona field*).

May
The Puerto La Cruz refinery goes on stream.

May 13
Cosutrapet, the small Communist federation of unions, is dissolved by decree, following an attempt to organise a general strike.

May 31
A significant discovery at Dación is made by Mene Grande (*Maturín basin, Oficina group, 20 km east of the East Guara field*).

June 25
Korean War starts.

June 30
The International Labour Office publishes its report on the conditions of work and freedom of association in Venezuela. The Military Junta sharply attacks the contents and conclusions of the report.

August
A 130-km water line from Siburúa springs to the refineries in the Paraguaná peninsula is completed by the Government, Creole and Shell.

October 12
The Güere oilfield is discovered by Sinclair on completion of wildcat Güere-1 (*Maturín basin, Oficina group, 40 km northwest of the Chimire field*).

October 15
Creole discovers the Alturitas field, on successful completion of Alturitas-1. At a total depth of 5208 m, this well is the deepest drilled outside the USA at the time (*Maracaibo basin, 140 km southwest of Maracaibo*).

October 21
The El Chaure refinery at Puerto La Cruz goes on stream.

November 13
Colonel C. Delgado-Chalbaud is assassinated. G. Suárez-Flamerich heads Government Junta.

November 14
Gas well Barbacoas-1 is completed by Atlantic (*Maturín basin, 30 km north of the Las Mercedes field*).

November 29
First public posting of oil prices, by Socony in the Middle East.

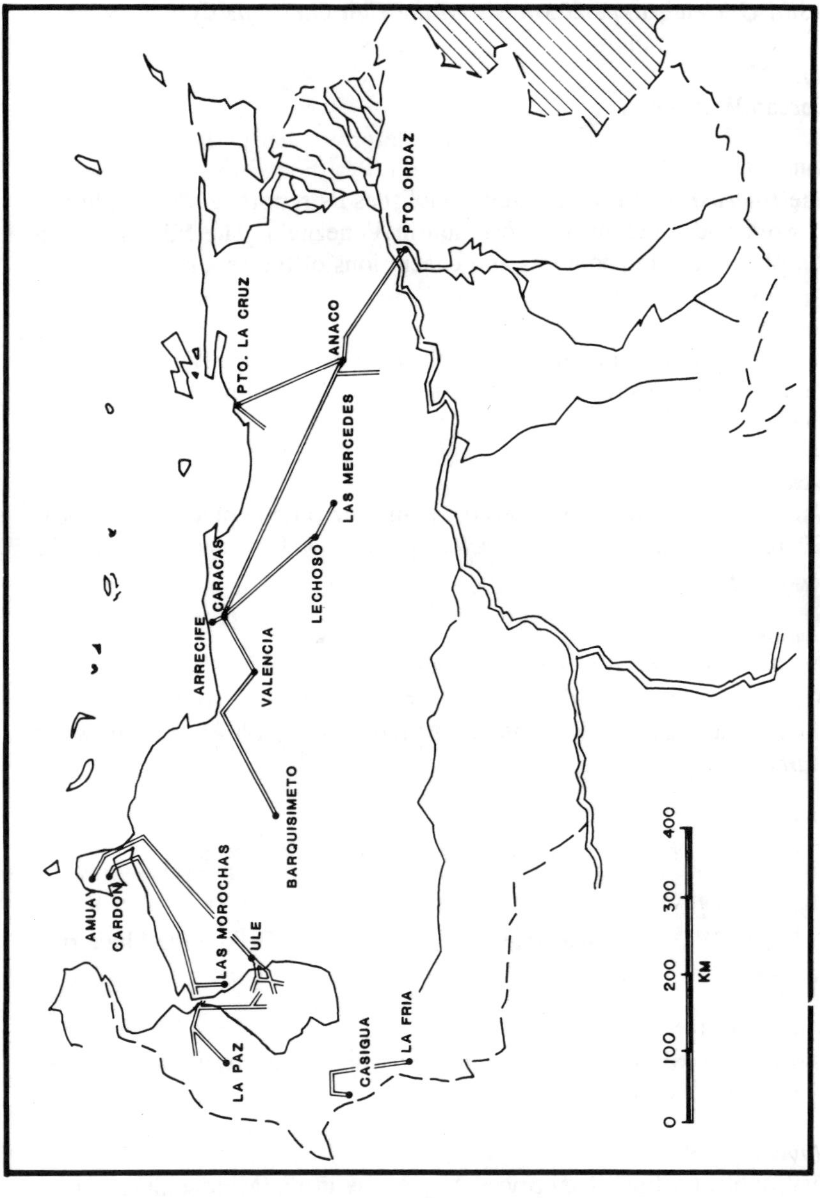

Fig. 15. Gasducts.

December 3
Resolution 3825 of the Ministry of Development declares that unitisation is compulsory for any reservoir being exploited by two or more operating companies.

December 29
By Royal Decree, the '50–50' principle profit-sharing is introduced in Saudi Arabia. It is the first country to follow the pattern established by Venezuela in the Income Tax Law of 1948.

December 30
Decree 41 creates the Ministry of Mines and Hydrocarbons. Santiago Vera-Izquierdo is the first minister.

1951

February 15
Mene Grande discovers the Cerro Pelado field on completion of Cerro Pelado-1A, renamed CPZ-1X (*Maturín basin, Anaco group, 30 km northeast of the Santa Rosa field*).

March 15
The Iranian Majlis agree to the nationalisation of Anglo-Iranian.

March 25
The giant Boca oilfield is discovered by Mene Grande on completion of new field wildcat SG-101X (*Maturín basin, Oficina group, 10 km north of the Chimire field*).

April 9
The Military Junta issues Decree 114 establishing temporary norms to regulate the working conditions in the oil industry, in place of the collective contract – negotiated by the unions – which had expired.

April 18
Atlantic discovers the Cocomón gasfield, on completion of new field wildcat Copa-1 (TUX-9) (*Maturín basin, 75 km east of the Las Mercedes field*).

May 27
The East Soto field is discovered by Mene Grande on completion of

new field wildcat SXM-201 (*Maturín basin, Oficina group, 5 km east of the Soto field*).

June 8
GG-301X (North San Tomé-1) is completed by Mene Grande as the discovery well of the Central Guara field (*Maturín basin, Oficina group, 20 km northeast of the Oficina field*).

June 28
New formulas to determine the value of Venezuelan crudes, for royalty payment purposes, are established.

July 2
CaM-301X is completed by Mene Grande as the discovery well of the Tascabaña field (*Maturín basin, Oficina group, 10 km west of the Chimire field*).

July 30
After 712 days total rig time, British Controlled Oilfields abandons new pool wildcat EM-285 at 4250 m total depth (*El Mene de Mauroa field*).

August 15
Creole and Mene Grande sign first unitisation agreement, relating to reservoirs of the West Guara field (*Maturín basin*).

September 3
Esquina-1R is successfully completed by Socony (*Maturín basin, Oficina group, 10 km southwest of the Chimire field*).

September 4
CaZ-401X is completed by Mene Grande as the discovery well of the East Caico field (*Maturín basin, Oficina group, 10 km northeast of the Caico Seco field*).

September 18
The First National Petroleum Convention ends.

September 22
Mata-2 is completed by Texas as a new field discovery well (*Maturín basin, 30 km northeast of the West Guara field*).

October 5
The Organic Statute of Compulsory Social Security Law and its general regulations are enacted.

October 20
Texas discovers the small Mata-1 field (*Maturín basin, Oficina group, 20 km north of West Guara*).

November 16
Collective Contract with marine officers is signed.

November 23
Collective Contract with marine workers is signed.

December 15
A small discovery is made by Mene Grande on completion of new field wildcat 1-MZX-1 (*Maracaibo basin, 25 km north of the La Paz field*).

December 24
Shell abandons new field wildcat Conquistado-1 (CQ-1); rig time was 838 days (*20 km west of Trujillo*).

December 25
New field wildcat Dakoa 8–1 is completed as the discovery well of the Dakoa field by Las Mercedes (*Maturín basin, 40 km east of the Las Mercedes field*).

December 31
A total of 171 crew-months are completed by geological surface parties.

The number of oil wells completed during any one year rise sharply and exceed 1000 for the first time.

1952

W.H. Bucher prepares first structural map of Venezuela, for the Geology Institute of the Ministry of Development.

L.G. Arcay and C. Romero-Zuloaga of Mene Grande publish an index of laws in force (*Indice de Leyes Vigentes*).

1952

January 16
An Executive Commission is established to co-ordinate work in connection with deepening of the Maracaibo Bar.

February 26
Shell completes the 'light line', a 72-cm, 255-km pipeline from Palmarejo de Mara to Paraguaná.

May 4
Creole completes new field wildcat Guanoco-2 as a discovery of the East Guanoco field (*Maturín basin, 20 km east-southeast of the Guanoco field*).

June 21
The Oscurote field is discovered by Socony on completion of well OV-1 (*Maturín basin, Oficina group, 25 km northeast of the West Guara field*).

June 27
C-151 is completed by Shell as a deeper pool discovery well in Cretaceous limestone (*La Concepción field*).

Autonomous institute of the Ministry of Mines and Hydrocarbons to co-ordinate work of dredging Maracaibo *bar* (*Instituto Nacional de Canalizaciones*), is established.

June 29
Phillips completes new field wildcat GC-1, discovering the Bella Vista oilfield (*Maturín basin, 55 km southeast of the Las Mercedes field*).

June 30
Atlantic completes the gaslines from the Placer, Lechozo, Punzón and Las Mercedes fields to the Caracas and Valencia industrial areas.

July 9
Wildcat Rincón 5–1 is successfully completed by Texas (*Maturín basin, Anaco group, 7 km east of the Santa Ana field*).

July 24
First postings of Venezuelan crudes. Creole publishes the prices at which it would sell Venezuelan oil in cargo lots at the deepwater loading terminals.

August 28
A revised trade agreement between Venezuela and the USA is signed in Caracas. There will be no more quotas for petroleum crude of 25° API or more, and products, exported to the USA, and the tariffs on hydrocarbons are again reduced by half for crudes of less than 25° API.

September
The gasline from the San Joaquín group of fields to Puerto La Cruz is inaugurated.

The general conservation policy for hydrocarbons is presented for the first time at a meeting of the Inter State Oil Compact Commission in Banff, Canada.

September 12
New field wildcat Mata 3–1 is successfully completed by Texas (*Maturín basin, Oficina group, 25 km northeast of West Guara field*).

OM-501X, formerly South Esquina-I, is completed by Mene Grande as a successful new field wildcat (*Maturín basin, Oficina group, 10 km northwest of the Oficina field*).

October 5
An important discovery is made by Creole in the Bolívar Coastal field, on completion of new pool wildcat TJ-319.

October 12
The revised trade agreement between Venezuela and the United States becomes effective. The quota system for petroleum imports into the US is abolished. Tariff on fuel oil is reduced by 50%.

October 28
TaG-5A is completed by Mene Grande as the discovery well of the Tácata field (*Maturín basin, 8 km west of Jusepín – Tacat field*).

November 12
Mene Grande, a wholly-owned subsidiary of Transocean Gulf (a subsidiary of Gulf), is rechartered in Delaware.

November 15
Wildcat CaZ-201X is completed by Mene Grande as the discovery well

of the Caracoles field (*Maturín basin, Oficina group, 5 km south of the Caico Seco field*).

November 19
Roblote-1 is completed by Sinclair as a gas producer (*Maturín basin, Anaco group, 25 km west of the El Toco field*).

November 28
LG-501X, formerly Lobo-1, is completed by Mene Grande as a successful new field wildcat (*Maturín basin, Oficina group, 30 km east of the East Guara field*).

November 29
A permanent Inter-Ministerial Commission of the Ministries of Finance and Mines and Hydrocarbons is established to study problems relating to the application of the Income Tax Law to the petroleum industry.

December 1
American interests purchase British Controlled Oilfields, one of the oldest operating companies in Venezuela. The syndicate operates under the name Talón Petroleum.

Wildcat Lejos-1 is completed by Mene Grande as a gas producer (*Maturín basin, Oficina group, 25 km east of the East Guara field*).

December 2
Colonel Marcos Pérez-Jiménez assumes dictatorial power.

December 22
Wildcat Sapo-1 is completed by Mene Grande as a gas producer (*Maturín basin, Oficina group, 3 km southeast of the Mapiri field*).

December 31
Mene Grande (incorporated in Delaware) acquires all assets and business of Mene Grande (a company incorporated in Venezuela).

Cumulative crude oil production exceeds 1000 million m^3. Production of crude oil during the year is over 100 million m^3 and that of natural gas 20 000 million m^3.

1953

January 13
The North Chimire field is discovered by Socony on completion of North Chimire-2, renamed NCHV-2 (*Maturín basin, Oficina group, 8 km northeast of the Chimire field*).

February 6
Venezuela becomes an associate member of the Inter State Oil Compact Commission.

February 12
New field wildcat NOV-1 is completed successfully by Socony as the discovery well of the North Oscurote field (*Maturín basin, Oficina group, 5 km north of the Oscurote field*).

February 23
Wildcat SM-501X (Siglo-1) is completed by Mene Grande as the discovery well of the North Soto field (*Maturín basin, Oficina group, 20 km north of the Chimire field*).

March 16
The Ministry of Mines and Hydrocarbons sets regulations for the pricing and transport of natural gas.

April
Prolific production from basement igneous and metamorphic rocks is discovered in the La Paz oilfield on deepening of well P-86. The level of production of the reservoir and the volume of the resources make this the largest reservoir ever found in non-sedimentary rocks in the world.

April 15
The Venezuelan Constituent Assembly promulgates a new Constitution.

April 21
The US Department of Justice terminates grand jury proceedings in the international cartel case and files civil suits.

May 7
Compañía Shell de Venezuela, a wholly-owned subsidiary of Bataafsche Petroleum Maatschappij, is incorporated in Toronto.

May 23
New field wildcat YS-401X (Yuca-1) of Mene Grande discovers the South Yopales field (*Maturín basin, Oficina group, 20 km south of the Oficina field*).

May 24
Sinclair makes the second discovery in the Barinas basin on completion of new field wildcat Sinco-1. This giant oilfield is the largest in the basin (*5 km west of the Silvestre Field*).

May 29
The Tiguaje field is discovered by Texas on completion of new field wildcat Tiguaje-1 (*Falcón basin, 130 km southwest of Coro*).

June 2
The Jobo oilfield is discovered by Creole on completion of wildcat JOM-2, formerly TY-42 (*Maturín basin, 25 km southwest of the Temblador field*).

June 11
New field wildcat LCV-1R is completed successfully by Socony, as the discovery well of the La Ceibita field (*Maturín basin, Oficina group, 15 km north of the Soto field*).

June 15
Tacat-3 is completed by Atlantic as the discovery well of the Tacat field (*Maturín basin, Jusepín group, 10 km west of the Santa Bárbara area, Jusepín field*).

Wildcat Rincon 7–1 is successfully completed by Texas as a discovery well (*Maturín basin, Anaco group, 10 km east of the Santa Ana field*).

August 20
New field wildcat NS-601X of Mene Grande discovers the Nigua field (*Maturín basin, Oficina group, 10 km north of the West Guara field*).

September 1
Well ZM-1X of Mene Grande discovers the Zorro field (*Maturín basin, Oficina group, 10 km north of the Zumo field*).

September 2
The Zeta-1 wildcat is completed by Mene Grande as a discovery well (*Maturín basin, Oficina group, 30 km north of the West Guara field*).

September 12
Collective Contract is signed with the oil companies by (unrepresentative) Independent Union Oil Committee (*Comité Petrolero Sindical Independiente*).

September 17
LS-1387, a deeper pool test of Shell in the Lagunillas area of the Bolívar Coastal field, is abandoned dry at a total depth of 5345 m, the deepest hole then drilled outside the USA.

October 24
New field wildcat OX-401X is completed by Mene Grande as the discovery well of the Oca field (*Maturín basin, Oficina group, 20 km southeast of the Oficina field*).

October 26
East Pilón-2 is completed by Creole as the discovery well of the Isleño field (*Maturín basin, 20 km southeast of the Temblador field*).

November 5
Collective Contract signed by the marine workers and oil companies.

November 14
Shell recompletes DM-22 as the first basement well of the Mara field.

December 31
Cumulative production of natural gas exceeds 200 000 million m^3.

1954

There is a new word in the oil industry: *venezolanisation.*

1954

January 22
Mene Grande discovers the Zumo field on completion of wildcat ZM-101X (*Maturín basin, Oficina group, 30 km northeast of the West Guara field*).

February 27
Collective Contract with marine officers signed by oil companies.

April 5
An important discovery is made by Mene Grande on completion of NZ-701X, formerly Nardo-1 (*Maturín basin, Oficina group, 20 km northeast of the West Guara field*).

April 8
New field wildcat Cagigal-1 is completed by Shell as a small producer (*Maturín basin, 120 km northeast of the Las Mercedes field, 80 km west of the San Joaquín field*).

April 24
The Araibel field is discovered by Socony on successful completion of wildcat ARV-1R (*Maturín basin, Oficina group, 25 km northeast of the Chimire field*).

June 24
Mahomal-1 is completed by Sinclair as a new condensate field discovery well (*Maturín basin, 90 km east-southeast of the Las Mercedes field*).

June 29
New pool wildcat VLA-14 is completed by Shell in the Paleocene Guasare formation (*Bolívar Coastal field*).

July 19
New field wildcat Elotes-1 (ES-1X) is successfully completed by Mene Grande (*Maturín basin, Oficina group, 50 km west of the Oficina field*).

July 20
Wildcat well LZ-601X of Mene Grande discovers the giant Limón oilfield (*Maturín basin, Oficina group, 15 km north of the East Guara field*).

July 24
BU-4, the fourth of the Burgúa wildcats drilled by Shell is abandoned (*Burgúa area: 55 km southeast of San Cristóbal, in the westernmost part of Apure State*).

July 28
Wildcat LG-701X of Mene Grande discovers the Lido oilfield (*Maturín basin, Oficina group, 15 km northeast of the East Guara field*).

September
School of Petroleum Engineering of Zulia University in Maracaibo starts functioning.

October 20
Wildcat Dakoa 10–1 is completed by Las Mercedes as an oil producer (*Maturín basin, 30 km east of the Las Mercedes field*).

October 29
An agreement is signed between Iran and a consortium of oil companies, thus ending the situation created by the nationalisation of the industry in 1951.

November 9
The Manresa field is discovered by Creole on completion of new field wildcat MR-1, formerly JX-8 (*Maturín basin, 25 km north of the Jusepín field*).

November 19
The West Nipa field is discovered by Socony on successful completion of new field wildcat W-NV-51R (*Maturín basin, Oficina group, 5 km west of the Nipa field*).

December 31
Casón-1 is successfully completed by Mene Grande (*Maturín basin, 30 km northwest of the Anaco group – El Tocó field*).

More than 11 million ha are explored by aerogeology.

The capacity of Venezuelan refineries exceeds 25 million tonnes crude oil per day for the first time.

Domestic consumption of petroleum products exceeds 5 million m^3 for the first time.

1955

A new structural map of Venezuela is published by the Geology Bureau, Ministry of Mines and Hydrocarbons.

February 1
Shell posts the prices of its Venezuelan crudes.

February 14
Creation of Venezuelan Geological Society.

April 10
Creole completes new field wildcat UD-1 (ex TJ-342), the discovery well of the Urdaneta giant oilfield (*Maracaibo basin, along western shores of Lake Maracaibo, 60 km south of Maracaibo*).

May 11
The Aníbal field is discovered by Atlantic (*Maturín basin, 40 km west of nearest Oficina field – Elotes*).

July 25
The Oleos wildcat (renamed OS-601X) is successfully completed by Mene Grande (*Maturín basin, Oficina group, 20 km east of the Oveja field*).

August 14
An important discovery is made by Mene Grande in the Oficina area of Eastern Venezuela on completion of wildcat MS-401X, formerly Melones-1 (*Maturín basin, Oficina group, 35 km southeast of the East Guara field*).

August 29
Minor amendment to the 1943 Hydrocarbons Law. Drilling of what really are exploratory wells, as if they were shallow holes dug for exploration surveys, is prohibited.

October 27
Sinclair makes a significant discovery on completion of wildcat Aguasay-1 (*Maturín basin, Oficina group, 40 km northeast of the West Guara field*).

Fig. 16. Manifold in Oficina field, Maturín basin.

December 20
The Zapatos giant field is discovered by Mene Grande on completion of wildcat ZM-301X (*Maturín basin, Oficina group, 40 km northeast of the Chimire field*).

December 27
Wildcat SM–150X of Mene Grande is completed as the discovery well of the Soyas field (*Maturín basin, Oficina group, 15 km north of the Chimire field*).

December 31
A record 171 exploratory wells are completed during the year.

1956

The Venezuelan stratigraphic lexicon is published.

January 11
Regulations of the Income Tax Law are issued.

1956

January 12
The Minister of Mines and Hydrocarbons, Edmundo Luongo Cabello, announces that concessions will be granted soon.

June 1
Decree 356 sets forth regulations for the use of the deep channel dredged in the Maracaibo *bar*.

June 4
AdM-301X, discovery well of the Adas field, is completed by Mene Grande (*Maturín basin, Oficina group, 15 km south of the Oritupano field*).

June 15
Phillips completes Gozo-2 as a gasfield discovery well (*Maturín basin, 50 km southeast of the Las Mercedes field*).

June 30
Decree 367 of the Ministry of Mines and Hydrocarbons establishes the Venezuelan Institute of Petrochemicals (*Instituto Venezolano de Petroquímica*).

July 11
Mene Grande discovers the North Zumo field on completion of ZM-201X (*Maturín basin, Oficina group, 5 km north of the Zumo field*).

Adrales-1 (AdM-401X) is completed by Mene Grande as a new field discovery well (*Maturín basin, Oficina group, 10 km west of the Oritupano field*).

July 27
Law relating to the Territorial Sea, Continental Shelf, Protection of Fisheries and Air Space is passed.

August 8
FZ-250X is completed by Mene Grande as the discovery well of the Finca field (*Maturín basin, Oficina group, 30 km northwest of the Oficina field*).

August 16
Concession cycle starts. Shell, Mene Grande, Superior and Venezuelan American Independent Oil Producers are granted 27 parcels covering 20 000 ha in Lake Maracaibo.

September 6
LM-50X (Leguas-1) is completed by Mene Grande as an oil producer (*Maturín basin, Oficina group, 12 km east of the East Guara field*).

October 6
Aguasay-3 is completed by Sinclair as a new field discovery (*Maturín basin, Oficina group, 45 km northeast of the West Guara field*).

October 11
Mene Grande completes as an oil producer well LG-550X, formerly Lustro-1 (*Maturín basin, Oficina group, 3 km south of the Lobo field*).

October 17
LM-301X (Lestes-1) is completed by Mene Grande as an oil producer (*Maturín basin, Oficina group, 5 km northeast of the Lobo field*).

October 19
Collective Contract signed between National Confederation of Workers (*Confederación Nacional de Trabajadores de Venezuela*) and operating oil companies.

October 26
Collective Contract with marine workers signed by oil companies.

December 6
Mene Grande receives 22 960 ha in 47 parcels in Lake Maracaibo. Last concessions granted during the year.

December 26
Ministry of Mines and Hydrocarbons fixes for the first time rates for ships using the new channel across Maracaibo *bar*. Majority of shipping involves oil tankers.

December 31
Number of oil-producing wells in the country for the first time exceed 10 000.

1957

Creole and Socony drill first exploration wells in the Gulf of Venezuela basin.

In Caracas, IBM installs first 650 electronic computer in Latin American.

January 21
A resolution of the Ministry of Mines and Hydrocarbons assigns to the Petrochemicals Institute the administration of a national network of gas pipelines.

March 14
One 9550-ha concession is granted to Venezuelan American, the first during the year.

April–May
Seven private companies conduct an aeromagnetic survey covering most of Apure and Barinas States.

April 22
Mene Grande completes the first producer in the Ceuta area of the Bolívar Coastal field.

May 20
Well MG-401X (Miga-1) is completed by Mene Grande as a new field discovery well (*Maturín basin, Oficina group, 30 km southeast of the Oficina field*).

June
The weekly newsletter (*Carta Semanal*) of the Ministry of Mines and Hydrocarbons starts publication.

June 26
Creole inaugurates the second huge offshore gas reinjection plant (TJ-2) in the Tía Juana area of the Bolívar Coastal field.

July 29
A voluntary programme for control of oil imports into the USA (east of the Rocky Mountains) is established. The US Secretary of Interior appoints a Special Assistant to administer the programme.

August 1
CL-1 is completed by Creole as the discovery well of a giant field (*Maracaibo basin, central area of Lake Maracaibo, 100 km south of Maracaibo, 30 km west of the Ceuta area of Bolívar Coastal field*).

September
First all-aluminium drilling platforms are installed offshore Lake Maracaibo for Superior.

September 7
Superior makes the first discovery in the concessions granted during 1956 and 1957: Lama-1 is completed as a successful new pool wildcat (*Bolívar Coastal field*).

September 15
Mene Grande completes well GM-850X (Grúas-1) as the discovery well of the North Guara field (*Maturín basin, Oficina area, 6 km north of the West Guara field*).

October 6
Large reservoirs of Eocene and Cretaceous (Mito Juan formation) are discovered by Sun on completion of new field discovery well SVSX-1, in its recently granted concession in Lake Maracaibo, south of Superior leases, Lama area (*Bolívar Coastal field*).

October 16
Superior starts production 14 months following the grant of titles.

October 20
Richmond discovers the Los Claros giant oilfield on completion of wildcat Zulia 22D-1 (*Maracaibo basin, 60 km southwest of Maracaibo*).

October 30
Creole receive 66 exploitation parcels covering 30 667 ha in Monagas State. This closes concession granting cycle. A total of 510 350 ha of exploration–exploitation concessions and 310 757 ha of exploitation concessions were granted in 1956 and 1957, at a cost of US$685 million. At least one-seventh of the volume of resources of the country were given to the concessionaires.

November 12
Signal successfully completes new field wildcat Centro-2X as the discovery well of the Centro oilfield (*Maracaibo basin, 80 km south of Maracaibo, 15 km south of the nearest production in Bolívar coastal field*).

November 20
New field wildcat Palmita-1 is completed by Socony as a producer. It is the first discovery in the Barinas basin in 4 years (*5 km north of the Silvestre field*).

December 2
Collective Contract signed by oil companies with marine officers.

December 12
The US Voluntary Oil Imports Programme is extended to cover the area west of the Rocky Mountains.

December 31
A record 415 million m^3 of crude oil were discovered during the year.
A total of 7·5 million ha are covered by seismic exploration.
A record 1739 wells were completed during the year.
Annual production of natural gas exceeds 30 000 million m^3 for the first time.

1958

January 23
Dictator Pérez-Jimenez is ousted by popular uprising; Rear Admiral W. Larrazábal heads Provisional Government Junta.

February 6
Venezuelan Petroleum becomes Sinclair Venezuelan Oil.

February 12
Mene Grande discovers the Yucal gasfield (*Maturín basin, 60 km north of the Las Mercedes field*).

February 15
The US Subsecretary of State for Latin America visits Venezuela regarding changes to the Imports Programme.

February 20
The Ipire-7 field is discovered by Las Mercedes (*Maturín basin, 120 km southeast of the Las Mercedes field, 120 km west of the Oficina group – Oficina field*).

March
Shell starts the first large-scale steam injection secondary recovery project in the world, in the Cabimas area of the Bolívar Coastal field.

A delegation headed by the US Undersecretary of State for Economic Affairs holds discussions in Caracas. The Ambassador of Canada also attends.

April 12
New field wildcat POSA 112–1 is first well successfully completed offshore in the Venezuelan portion of the Gulf of Paria. A total of seven wells were drilled later in the field, which has not been put on production (*30 km northeast of the Pedernales field*).

May 20
Report of an investigating commission is submitted to the Government Junta concerning the concession granting programme of 1956–7. All legal procedures were found indisputable.

Fig. 17. Puerto Miranda terminal.

May 22
New field wildcat 6Y-2X of Mene Grande is completed as the discovery well of the North Lamar field (*Maracaibo basin, offshore in Lake Maracaibo, 12 km northeast of the Lamar field, 15 km southeast of the nearest area of Bolívar Coastal field – Lama*).

June 4
The United States issues Regulations relating to imports of unfinished oils.

July 1
Transport, storage and installation of systems for handling of liquid petroleum gases can only be carried on with a permit from the Ministry of Mines and Hydrocarbons.

July 10
The Income Tax Law is amended, providing for a general increase in the level of taxation.

August 8
Mene Grande discovers the Lago field on successful completion of new field wildcat 2Y–2X (*Maracaibo basin, offshore in Lake Maracaibo, 100 km south of Maracaibo, 10 km east of the Lama area of the Bolívar Coastal field*).

August 12
The (Finance and Mines and Hydrocarbons) Inter-Ministerial Commission is to study problems relating to refining and transportation in the petroleum industry.

September 10
The US Oil Import Administration proposes import quotas be given only to domestic refiners (so-called Carson plan).

October 3
The Morichal giant oilfield is discovered by Phillips on completion of new field wildcat MPG-4-1 (*Maturín basin, 40 km southwest of the Temblador field*).

November 27
New field wildcat LPG 114-3 of Phillips discovers the Lamar giant field, in new concessions given the previous year offshore in Lake Maracaibo (*Maracaibo basin, 110 km south of Maracaibo, 30 km south of the Lama area of the Bolívar Coastal field*).

December 17
The US Undersecretary of State for Economic Affairs again visits Venezuela for discussions on the discriminatory provisions in the imports programme.

December 18
Presidential elections. Rómulo Betancourt is elected.

December 19
Decree 476 of Provisional President Edgar Sanabria introduces changes in the Income Tax Law, increasing tax rates to a maximum of 46% of net earnings, in addition to the 'cedular' income tax of 1·5%. When combined with royalties and other tax payments, the Venezuelan Government revenue increases to 66–67% of the industry's gross profits.

December 23
The Law for practising Engineering and Similar Professions (*Ley del Ejercicio de la Ingeniería y Profesiones afines*) is sanctioned by Provisional Presidential Sanabria.

1959

Shell starts *in situ* combustion pilot project in area east of Tía Juana (*Bolívar Coastal field*).

January 16
The post of Petroleum Counsellor is established in the Venezuelan Embassy in Canada.

February 6
Shell starts a general reduction of the posted prices of Venezuelan oils.
President of Creole is replaced. He had sharply criticised in a letter

the amendment to the Income Tax Law; Mines and Hydrocarbons Minister, Julio Diez, flatly rejected the contents of the letter.

February 12
Reductions of even larger amounts in posted prices are introduced in the Middle East; posted prices in the USA remain unchanged, making the breakdown of the international crude oil pricing structure evident.

March 10
Proclamation 3279 by US President Eisenhower establishes mandatory control of oil imports into the United States. Imports are considered to be a threat to national security.

April 4
Posted prices of Venezuelan crude oils are uniformly reduced by operating companies.

April 9
The Coordinating Commission for the Conservation and Commerce of Hydrocarbons is created by Resolution 557 of the Ministry of Mines and Hydrocarbons. The Commission shall be 'capable of studying and recommending the regulations on the commerce of hydrocarbons and of coordinating them with the conservation policy advised by the supreme interest of the nation'.

April 20
Secret consultations take place in Cairo, at the time of the First Arab Petroleum Congress, between the delegates of Venezuela, Iran, United Arab Republic and Kuwait, and the Head of the Petroleum Department of the Arab League. This meeting set up the general agreement on the establishment of an 'oil consultation commission', later to be known as OPEC.

May 13
Minister Pérez Alfonzo explains the Venezuelan position with respect to discriminatory provisions in the imports programme to the US Secretaries of Commerce, State and Interior.

May 14
Budare-1 is completed by Socony as an oil producer (*Maturín basin, Oficina group, 15 km southwest of the Elotes field*).

May 25
The Minister of Foreign Affairs insists in a letter to US Ambassador that consultations at high levels are necessary to preserve the price of Venezuelan oils on the international markets, in view of the introduction of a programme of crude imports in the USA.

May 27
Minister Pérez Alfonzo, addressing the Chamber of Deputies of the National Congress, gives a summary of the national petroleum policy.

July 7
Ministry of Mines and Hydrocarbons decree 850 establishes a Commission to plan the formation of petroleum technicians.

July 24
Shell completes CR-2 as the discovery well in the (El) Rosario field (*Maracaibo basin, 190 km southwest of Maracaibo, 50 km north of the Tarra group – Los Manueles field*).

August 18
Sun's 16 000 m^3/day automatic flow station on Lake Maracaibo goes on stream.

August 21
Icón-1 (IZ-104X) is completed by Mene Grande as a gas producer (*Maturín basin, Oficina group, 15 km southwest of the Caico Seco field*).

August 22
Law establishing National Institute for Educational Cooperation (*Instituto Nacional de Cooperación Educativa – INCE*), to provide a programme of skilled training and apprenticeship.

September 11
The National Energy Council (*Consejo Nacional de Energía*) is established.

October
Establishment of the first 'petroleum' Embassy, in Cairo, with jurisdiction in the UAR, Saudi Arabia and Iraq. Antonio Martín Araujo is appointed first Ambassador.

October 14
Venezuela attends the seventh meeting of the Committee of Arab Petroleum Experts in Jeddah (Saudi Arabia) as an observer. Dr Eduardo Acosta, the Venezuelan representative, stated unequivocally that 'only with the cooperation of the Middle East exporting countries' could oil prices be maintained.

November 29
The Third Venezuelan Geological Congress ends in Caracas.

December
The Treasury borrows Bs300 million from a group of US commercial banks, against future tax collections. Loan repaid in March and May.

Sinclair discovers the East Aguasay field on completion of new field wildcat E. Aguasay-2 (*Maturín basin, Oficina group, 20 km east of the Aguasay field*).

December 3
The Ministry of Mines and Hydrocarbons (*Oficio circular* 3825) makes compulsory unitisation of reservoirs exploited by more than one concessionaire.

December 31
Cumulative production of crude oil reaches 2000 million m^3.

1960

February
The Venezuelan College of Engineers establishes an Office of Control for Authorisations, to issue permits to foreign graduates (mostly geologists and petroleum engineers), as per provisions of the 1958 Law for practising the engineering profession.

February 13
Fedepetrol and oil companies sign new Collective Contract.

March 17
Unitisation agreement for Lower Lagunillas reservoir signed by Shell, Creole, Mene Grande and San Jacinto (*Bolívar Coastal field*).

April
Publication of *PODE* (Petroleum and other statistical data) starts.

April 5
Collective Contract signed by oil companies with marine workers union.

April 19
The Venezuelan State oil company, *Corporación Venezolana del Petróleo* (CVP) is established by decree 260 of President Betancourt.

April 20
Paria Operations discovers second offshore oilfield in the Venezuelan portion of the Gulf of Paria, on completion of new field wildcat POSA 117-1 (*Maturín basin, 25 km northeast of the Pedernales field*).

May 3
In a speech in Dallas at the annual TIPRO meeting, Minister Pérez Alfonzo reaffirms Venezuelan views on oil production programming, exports and prices.

August 9
Resolution 994 of the Ministry of Mines and Hydrocarbons informs concessionaires that production of oil to be sold at what the Coordinating Commission considers 'abnormal discounts', will not be permitted.

Esso reduces, by as much as US$0.14 per barrel, the postings of its crudes in the Persian Gulf.

August 12
Minister of Petroleum Affairs of Saudi Arabia, Abdallah Tariki, and Director General of the Arab League Petroleum Department, Mohammed Salman, approach Minister Pérez Alfonzo, concerning the reductions of postings in the prices of Middle East oil.

In accordance with Resolution 994, production of some companies is ordered closed.

August 15
Shell also reduces posted prices in the Persian Gulf (crudes from Iraq and Qatar), following Esso reductions.

August 16
British Petroleum brings down posted prices of its crudes by no more than US$0.10 per barrel.

August 20
Creole successfully completes wildcat Acema-1 (*Maturín basin, Oficina group, 20 km east of the Aguasay field*).

September 3
The Government of Iraq invites delegations at the highest level from Iran, Kuwait, Saudi Arabia and Venezuela, to meet in one week in Baghdad, in connection with the reductions in posted prices of Middle East oils started 9 August.

September 10
Socony announces a reduction in the posted prices of its Persian Gulf crudes to BP-indicated levels.

A meeting of representatives of Iran, Kuwait, Saudi Arabia and Venezuela convenes in Baghdad.

September 14
Esso announces an increase of US$0.04 per barrel from its Persian Gulf crudes.

OPEC (the Organisation of Petroleum Exporting Countries) is established, following a 4-day meeting in Baghdad between Iran, Iraq, Kuwait, Saudi Arabia and Venezuela. The principal aim of the Organisation shall be 'the unification of petroleum policies among Member Countries and the determination of the best means for safeguarding the interests of Member Countries individually and collectively' (para. 4 of Resolution I.2).

October 22
A strong Venezuelan delegation attends the Second Arab Petroleum Congress in Beirut.

December 4
A subcommittee meets in Baghdad to discuss the budget, statutes and headquarters of OPEC.

December 27
First directorate of the CVP is named.

December 31
The volume of estimated proved reserves stands at 2767 million m^3, an all-time high.

Wells completed during the year drop to 444.

The capacity of Venezuelan refineries exceeds 50 million ton/year for the first time.

The area covered by seismic exploration is drastically reduced to 114 ha.

1961

The El Palito refinery of Mobil goes on stream.

Sun drills a development well in the SVS area of the Bolívar Coastal field which has 562 m of net oil sand, the thickest producing section found in Venezuela.

January 4
Ministry of Mines and Hydrocarbons Resolution 2 assigns 140 000 ha to the CVP.

January 21
The Second Conference of OPEC ends in Caracas. Thirteen Resolutions are adopted, notably on prices of crude oil on the international market. Qatar is accepted as the sixth Member, the Statutes are approved and Geneva is selected as headquarters of the Secretariat.

January 23
The Constitution is adopted. The operation and administration of hydrocarbons are within the competence of the National Power; the State shall attend to the conservation of natural resources and their exploitation for the chief benefit of the Venezuelan people; 'under no circumstances' can new concessions be granted, unless authorised by the national congress.

February 13
Income Tax Law is amended. Payments by oil industry are placed on a 'pay as you go' basis.

1961

February 25
Alfredo Tarre-Murzi is appointed as first Governor for Venezuela to OPEC.

March 19
By Decree 390, exchange controls are introduced in Venezuela. The oil dollar rate continues at US$1 equals Bs3.09.

April 24
The national network of gas pipelines and its dependencies, rights and possessions are transferred by the Venezuelan Petrochemical Institute to the CVP.

May 4
The first session of the Board of Governors of OPEC ends in Geneva. The Board meets four times a year.

May 26
The National Congress promulgates Law agreeing to the treaty signed in Baghdad establishing OPEC.

June
Fedepetrol joins the International Federation of Petroleum and Chemical Workers.

June 26–29
A 'Conference of State Oil Companies on the Americas', invited by Minister Pérez Alfonzo, ends in Maracay.

June 29
The Law of Urgent Economic Measures is approved by the National Congress.

June 30
Income Tax Law is amended.

CVP completes its first well, in the Lamar field.

Collective Contract with marine officers union signed by oil companies.

July 19
Resolution 551 of the Ministry of Mines and Hydrocarbons assigns the national network of gasducts to the CVP.

August 15
Venezuela ratifies the International Convention of the Continental Shelf.

October 3
The Ministry of Mines and Hydrocarbons eliminates the regulation prohibiting drilling within 1200 m of national reserves, free zones and 'sobrantes'.

November 1
The Third Conference of OPEC ends in Teheran. Twelve Resolutions are approved, notably on the establishment of an Information Centre within the Secretariat and on a feasibility study of the fields within which, as far as practicable, uniformity in the practices, methods and techniques is desirable and convenient.

November 7
After 3 years of unsuccessful wildcatting in the Barinas basin, a new field is discovered on completion of Hato-1 by Socony (*8 km southwest of the Silvestre field*).

December 18
President Betancourt and US President J.F. Kennedy reaffirm in Caracas policy on Latin American basic resources.

December 31
For the first time in the development of the oil industry, there is in Venezuela a negative rate of increase of proved reserves, in the order of 83 million m^3 which cause a reduction in the volume of proved reserves.

Total crew-months of geological surface parties is drastically reduced to 12.

1962

February 24
The Casigua-La Fría gas pipeline of the CVP is inaugurated.

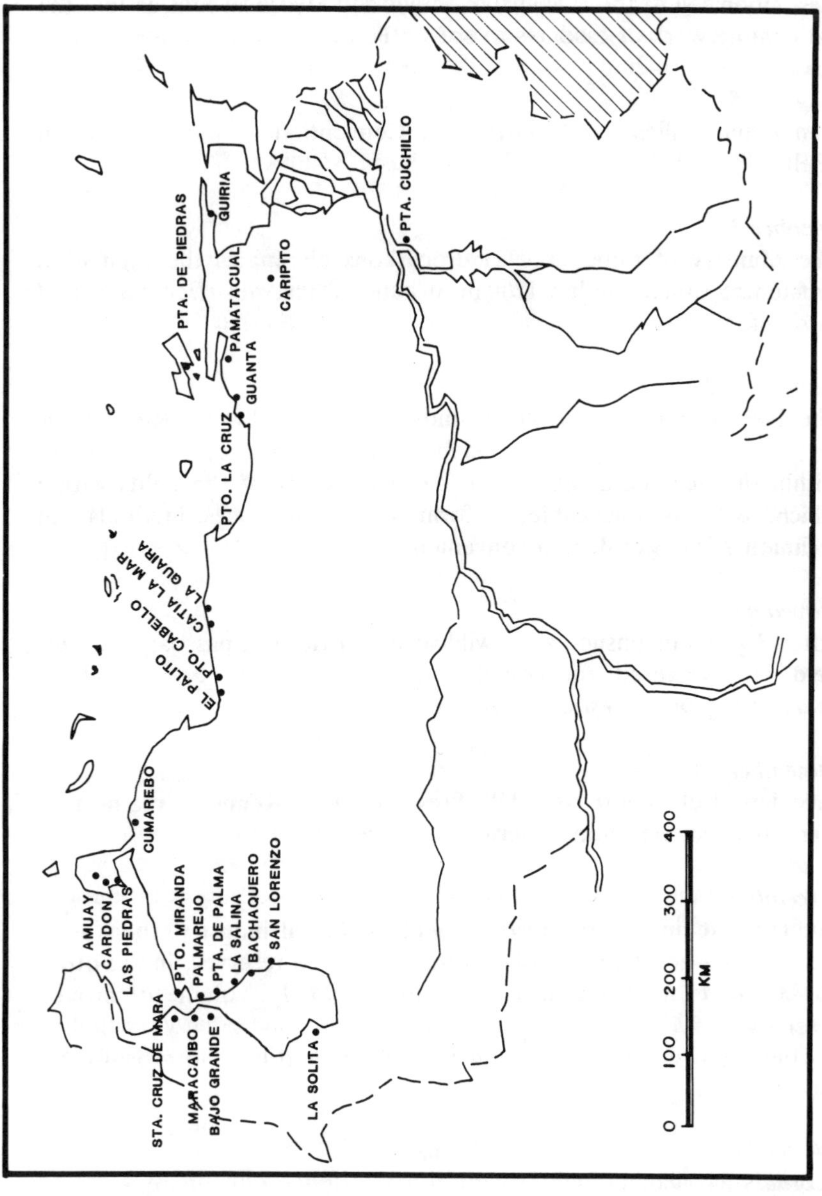

Fig. 18. Oil ports and terminals.

March 2
In a circular letter, the Ministry of Mines and Hydrocarbons reminds oil companies that if oil reservoirs extend beyond the limits of their leases, the petroleum resources therefrom are inappropriable by them.

March 19
13R-1, first CVP well in the Barinas basin, spuds in.

March 24
President Betancourt, in an address opening the First Venezuelan Petroleum Congress, says private oil companies will not be nationalised, but rather will work side by side with the CVP.

April 8
The First Session of the Fourth Conference of OPEC ends in Geneva.

May 11
Mobil completes Barquis-1A as a successful new field wildcat. This is the only new field discovery in Venezuela during the year (*Maturín basin, 50 km east of the Temblador field, 30 km west of the Tucupita field*).

June 4
Indonesia and Libya join OPEC.

June 8
The Fourth Conference of OPEC ends in Geneva. Three very important Resolutions are agreed upon concerning the restoration of posted prices of oil to prior-August 1960 levels, the expensing of royalties in the Middle East and the elimination of any contributions to the marketing expenses of the companies.

September 5
The Ministry of Mines and Hydrocarbons announces that gas supplied by one company to another, for reinjection, will not pay exploitation taxes.

November
Lagogás-I is inaugurated. It will reinject 4 million m^3 of gas a day to reservoirs of the Lagunillas area of the Bolívar Coastal field.

November 25
The First Session of the Fifth Conference of OPEC ends in Riyadh. Three Resolutions on administrative matters are approved.

November 30
US President Kennedy amends mandatory oil imports programme, fixing quotas on basis of production and not of demand. Exemptions for volumes imported 'overland' from Canada and Mexico are introduced.

December 4
Decree 906 provides that donations in excess of Bs5000 from oil companies must be approved by the Ministry of Finance to be deductible for income tax purposes.

December 12
An Assistant of US President Kennedy and the Subsecretary of Interior travel to Caracas to explain aspects of the imports programme.

December 31
Production for the year exceeds 500 000 m^3.

1963

Shell completes new pool wildcat VLE-400 as the first Cretaceous producer in Lake Maracaibo (*Lamar field*).

January 1
President Betancourt outlines in his New Year Message the programmes to be followed in the development of the national economy and in particular of petroleum policies.

January 3
Petroleum Counsellors are appointed to the recently established Embassies in Saudi Arabia and Iran.

January 4
The Venezuelan Petroleum Corporation (CVP) starts production from oil fields in the Barinas State.

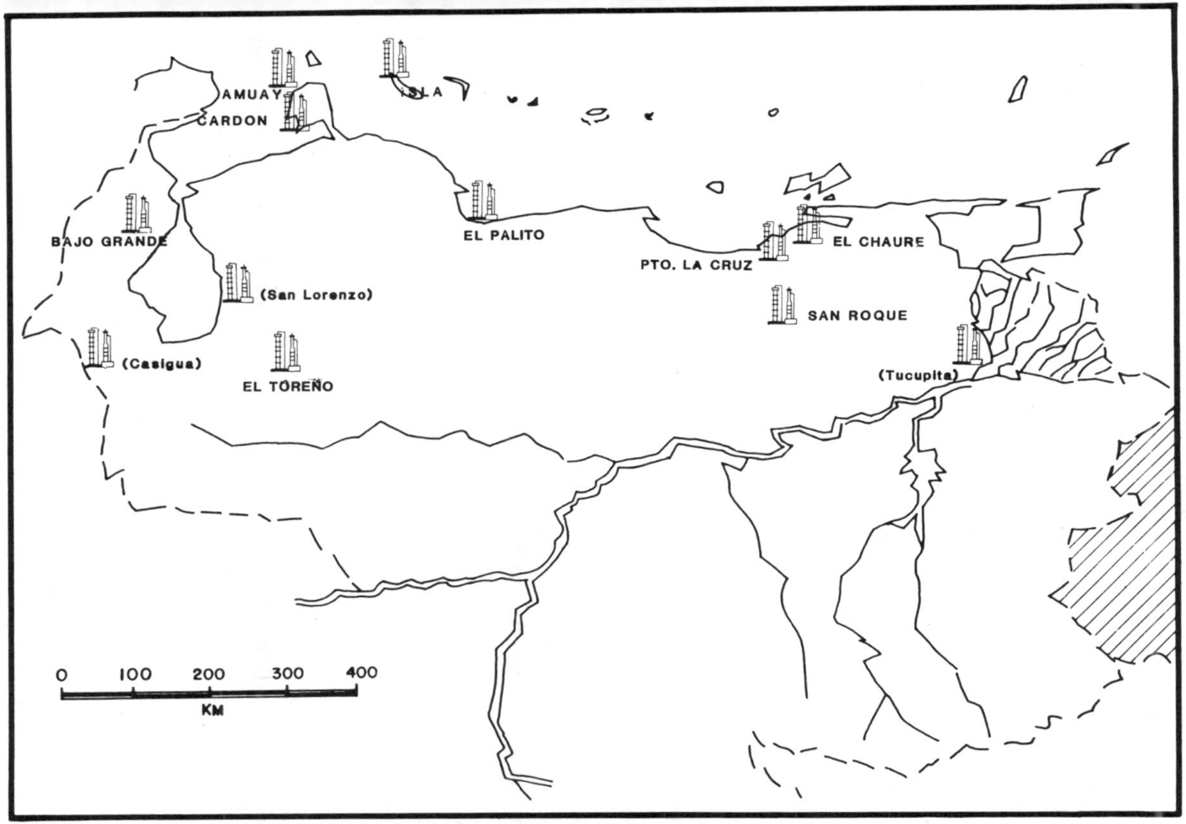

Fig. 19. Refineries.

1963

February 12
A 4-day first meeting of a joint Venezuela–US technical consultative group on the US imports programme ends in Washington.

February 20
Presidents Betancourt of Venezuela and Kennedy of the USA have an interview in Washington. There is agreement that Venezuela's interests should be acknowledged in any US oil programme.

March 15
The USA increases by 10% the Venezuelan residual oil quota.

March 29
Richmond completes Zulia 29F-1 as a small producer of heavy oil (*Maracaibo basin, 5 km southeast of the Los Claros field, 5 km northwest of the Urdaneta field*).

April 4
A Consultative Meeting of OPEC is held in Geneva.

May 7
A nationwide TV debate takes place on oil matters between Minister Pérez Alfonzo and Arturo Uslar-Pietri.

May 18
A new Collective Contract is concluded between Fedepetrol and representatives of the operating companies.

May 30
Collective Contract signed between marine workers union, represented by Federation of Workers in the Industry of Hydrocarbons and Derivatives (*Fetrahidrocarburos*), and operating oil companies.

June 5
The Chamber of Deputies declares that the import policy of the USA seriously prejudices Venezuelan oil.

July 1
The National Government takes over maintenance of navigation south of Cabimas.

July 4
The Second Consultative Meeting of Heads of Delegations of OPEC ends in Geneva. Algeria, Colombia, Nigeria and Trinidad attend as observers.

September 30
CVP sets up headquarters in Maracaibo. This move is considered of great significance for the economy of Zulia State.

October 9
The Third Consultative Meeting of OPEC ends in Geneva.

December 1
Orderly elections, a resounding vindication of the democratic process, are held; 95% go to the polls. Raúl Leoni is elected to succeed Betancourt.

December 4
The Fourth Consultative Meeting of OPEC takes place in Beirut.

December 5
The Supreme Court decides in favour of Mene Grande, an important legal issue concerning the disposal by the oil company of permanent installations which, at the end of the duration of a concession, would have passed free to the Government.

December 16
Manuel Pérez-Guerrero is appointed Minister of Mines and Hydrocarbons.

December 19
Minister Pérez-Guerrero announces that the Government will continue the oil policies of former Minister Pérez Alfonzo, although new and different modalities might be introduced in its application.

December 31
The Second Session of the Fifth OPEC Conference ends in Riyadh. A committee is established to continue negotiations with the oil companies concerning the expensing of royalties. Studies on a project for the creation of an economic commission on crude oil prices should be prepared.

The volume of cumulative production (2753 million m^3) exceeds the volume of proved reserves (2700 million m^3) for the first time.

Cumulative production of natural gas exceeds 500 000 million m^3.

1964

January 14
Texaco Maracaibo takes over the properties of Superior Oil of Venezuela.

January 18
The exchange control restrictions of 1961 are eliminated and a new rate for the oil dollar is fixed at Bs4.40 per dollar.

January 30
The Ministry of Mines and Hydrocarbons announces that concessionaire oil companies will not be authorised to open new service stations.

March 3
Shell Química de Venezuela, a Venezuelan corporation, is established.

March 11
In an inaugural address, President Leoni reaffims his regard for present concessions and the fact that service contracts will be the basis for future development.

March 31
The Morón refinery – one of the smallest in the country – is transferred by the Petrochemical Institute to the CVP.

April 6
The Tanker *Esso Maracaibo,* out of control due to an electric failure, collides with a section of the new 8-km concrete bridge over Lake Maracaibo. Some 300 m of roadway fall into the water; seven people die.

April 9
The Fifth Consultative Meeting of OPEC ends in Geneva.

June 12
Closing the Fedecámaras meeting, President Leoni asserts that national oil policies are positive and well-defined and that the CVP will promote oil activities.

July 14
The Sixth Conference of OPEC ends in Geneva. Five Departments are established in the Secretariat (Administration, Economics, Enforcement, Public Relations and Technical) and a procedure for final negotiations with the oil companies on the question of expensing of royalties is agreed upon.

July 17
Minister Pérez-Guerrero visits Canada.

August 13
A Collective Contract, similar to that signed by private oil companies, is signed between the CVP and Fedepetrol.

October
Seven companies join in a vast geophysical survey of the Gulf of Venezuela offshore area.

October 4
The first session of a special Consultative Meeting of OPEC is held in Beirut from 30 September to 4 October.

October 10
Lake Maracaibo bridge reopens to traffic, following partial destruction by Creole tanker.

October 28
A meeting of State oil companies of Latin America ends in Buenos Aires.

November 3
Presidential Decree 187 sets out a new system for regulating the distribution of petroleum products on the domestic market. The CVP will receive one-third of the market by the end of 1968. A quota will be established for the building of gasoline stations, based on estimated

consumption. Some of the stations presently held by private companies will also be transferred to CVP.

The second session of a special Consultative Meeting of OPEC ends in Geneva.

November 5
President Leoni, at the occasion of the formation of a coalition with the URD and FND political parties, says that service contracts are under discussion.

November 13
The third meeting of the technical Venezuelan-US committee studying the import programme problems ends.

November 28
The Seventh OPEC Conference ends in Djakarta. Five Resolutions are passed, one of them in partial settlement of the question of expensing of royalties and another establishing an economic commission for the study of prices on the international markets.

December 2
Resolution 1010 of the Ministry of Mines and Hydrocarbons sets at 107 655 m^3 the estimated demand of petroleum products on the domestic market, in connection with Decree 187.

December 8
The US Department of Interior announces minor changes in oil import regulations.

1965

February 2
The Organizational Meeting of ARPEL (Latin American State Oil Companies Association) ends in Lima. CVP is a member.

February 10
Amendments to oil import regulations concerning emergency advances for residual fuel oil quotas are put into force by the US Secretary of Interior.

Fig. 20. Morón petrochemical complex.

March 19
A 3-day meeting at the political level of Venezuelan and US officials ends. The problems of the imports programme are discussed.

July 1
Richmond becomes Chevron Oil of Venezuela.

July 13
The Ninth OPEC Conference ends in Tripoli, Libya. Resolution 61 establishes, as a temporary measure for the stabilisation of prices, a production programme among member countries.

September 1
OPEC headquarters moves to Vienna.

October 3
The First Ordinary Conference of ARPEL ends in Rio de Janeiro. The Constitution is approved. CVP and all Latin American State oil companies (except Pemex) become founder-members.

November 26
Establishment of Petrolera Mito Juan, a Venezuelan service company.

December 10
US President Lyndon Johnson sets mandatory oil import quotas for 1966. The 'special position' of Venezuelan oil for 'Western Hemisphere security' is recognized. Allocations are granted for petrochemical plants feedstocks and also for petrochemical facilities in Puerto Rico.

December 17
The Tenth OPEC Conference ends in Vienna. It is decided that posted, or reference, prices, should be used as the basis for the calculation of income tax liabilities in member countries and to give full support to Libya, in connection with the attitude of certain oil companies towards the recent amendment of the Petroleum Law of that country.

December 30
The Ministry of Mines and Hydrocarbons decides (*oficio* 3299) that from 1 January 1966, no discounts in excess of 10% below the posted prices of residual fuel oil for export, shall be authorised.

December 31
The volume of crude petroleum produced exceeds 200 million m^3. Cumulative production since inception is over 3000 million m^3.

More than 60 million tonnes of crude are processed during the year in Venezuelan refineries.

The area under concession drops to less than 3 million ha.

1966

January 5
Production during the current week averages 620 000 m^3/day.

March 11
The Ministry of Mines and Hydrocarbons notifies oil companies that effective from 1 April, the maximum permissible discount on fuel oil prices will be 15% off posted prices.

March 23
The Ministry of Mines and Hydrocarbons issues a Resolution requiring the private oil companies to cede to the CVP, prior to the end of the year, a sufficient number of outlets to represent at least 10% of the companies' 1964 domestic gasoline sales.

March 30
The US Department of the Interior issues regulations liberalising the US residual import programme. New quotas will be set at a total requested by marketeers.

April 1
Luis Alcalá Sucre takes over as President of Mene Grande, the first time a Venezuelan heads a concessionaire company.

April 5
Pancoastal sells its holdings in Venezuela to Texas.

April 14
Unigás-I, giant compression plant in Lake Maracaibo, goes on stream.

April 18
The US Administrator of Oil Import Administration issues Residual Fuel Oil Imports Instructions, in accordance with Oil Import Regulations issued on 30 March by the US Secretary of Interior.

April 28
The Eleventh Conference of OPEC concludes in Vienna. Reference prices shall be the base for income tax payments.

May 18
An agreement is signed between the CVP and the Municipal Council of Caracas, concerning the installations and rights for natural gas distribution within the city.

May 23
New York City issues a regulation establishing the maximum levels of sulphur content in heavy fuel oil used in the area at 2·2% to April 1969, 2·0% to April 1971, and 1% as from April 1971.

May 25
Following similar action on the part of the Chamber of Deputies, the Venezuelan Senate gives support to the national petroleum policies; in particular, the defence of prices in the international market, the tax claims, OPEC, Decree 187, fuel oil discounts limit and the service contracts approach.

May 28
President Leoni inaugurates the CVP Morón-Barquisimento 147-km gasline.

June 19
A meeting of the National Development Council is held to discuss the social problem of 'abandoned areas', where exploitation has gone to very low levels.

June 20
The National Energy Council is reconstituted and meets for the first time in years. It is composed of 18 representatives of government, private industry, trade unions and political parties.

June 27
A meeting of Fedecámaras approves recommendations asking the Government for revision of the oil policy, to allow mixed enterprises for development of petrochemicals, to spell out the basis of service contracts as soon as possible.

July 11
President Leoni signs the law approving the Treaty of Montevideo, indicating the intention of Venezuela to enter the Latin America Free Trade Association (LAFTA).

Law of the Venezuelan Workers Bank is enacted.

July 25
Second Ordinary Assembly of ARPEL ends in Lima.

July 29
Resolution 860 of the Ministry of Mines and Hydrocarbons orders the transfer of an additional 10% of domestic gasoline sale outlets by private companies to CVP during 1966, in 12 States and the Federal District.

August 9
By Decree 598, a Presidential Commission of 11 members, representing public and private sectors, is established to evaluate the present public expenditure system and, in particular, to study the function of oil revenues in the economic and social development of the country.

August 11
Collective Contract signed between Fedepetrol and major operating companies, providing for pay increases of between 10 and 12% and 99% stability.

September 1
The Chamber of Deputies approves a vague proposal to increase the taxes on the oil companies and to give the CVP all the domestic gasoline market.

September 8
CVP begins construction of a new gas distribution network for Caracas.

October–December
The El Toreño refinery in Barinas goes on stream.

October 4
The Technical Office of the Ministry of Mines and Hydrocarbons sets definitions and regulations on reserves of oil estimates.

October 6
Minister Pérez-Guerrero announces on a TV and radio national hook-up, terms of agreements with private oil companies on tax claims, introduction of reference prices and new income tax rates.

October 17
First Meeting of Representatives of OPEC National Oil Companies ends in Maracaibo. Six recommendations are approved, one of which refers to the study of ways and means of co-ordinating the entrance of the State companies into the world petroleum markets.

October 27
National Energy Council presents to Congress the basis and norms for service contracts.

November
Creole starts operation of its Lagunillas crude dehydration plant, the world's largest.

November 25
The President of Fedepetrol, Luis Tovar, is appointed President of the International Federation of Petroleum and Chemical Workers at the end of the Fifth World Petroleum Congress meeting in Caracas.

December 8
The Twelfth Conference of OPEC ends in Kuwait.

December 9
Venezuelan Congress approves tax bill which includes tax reform for private oil companies.

December 12
Shell Mara, the largest tanker flying the Venezuelan tricolour, is christened at Puerto Miranda.

The US Oil Import Administration amends the oil imports regulation defining refinery inputs.

The Second Ordinary Meeting of ARPEL ends in Montevideo.

December 23
New Income Tax Law (*Ley de Impuesto sobre la Renta*). The 'cedular' tax is eliminated and a new rate of 52% income tax is established for net incomes in excess of Bs28 million. For certain reinvestments in exploration, secondary recovery projects and conservation of gas, reductions of at the most 4% are possible. The effective date of the Law is 1 January 1967. It introduces the system of calculating tax liabilities on the basis of reference prices, which increase gradually over a period of 5 years to predetermined levels.

December 29
CVP inaugurates its 100th service station.

1967

January 3
José Antonio Mayobre is sworn in as Minister of Mines and Hydrocarbons.

January 9
Venezuela adheres to the Montevideo Treaty, LAFTA (Latin American Free Trade Association).

January 19
The New York–New Jersey Inter-State Air Pollution Abatement Conference recommends that fuel oil containing more than 1% sulphur by weight should not be used as of 1 October 1969 by steam or electric power companies, nor heating oil containing more than 0·3% sulphur by weight.

February 3
The Minister of Finance, E. Morales-Crespo, signs with Shell and Creole the agreements negotiated by Mines Minister, Pérez-Guerrero, introducing reference prices for the next 5 years and accepting income tax of 51% (formerly 47·5%). Also agreed, were tax claims for previous years: US $85 million for Creole and US $55·5 million for Shell, payable one-third on signature, one-third in 9 months and one-third in 18 months.

February 21
Mito Juan sets up a subsidiary to acquire two small blocks of concessions in the Maturín basin, southeast of Oficina.

February 24
A Seminar on the Oil Industry in Latin America, sponsored by the Economic Commission for Latin America, ends in Santiago, Chile. Representatives of Venezuela submit papers on hydrocarbon reserves, conservation, prices of oil in the international markets and financing.

April 8
During the Seventh World Petroleum Congress in Mexico City, the CVP presents three papers on the geology of the unexplored Gulf of Venezuela basin, the hydrocarbon resources of the tar belt of the Maturin basin and the exploration drilling programme at Guanarito.

Fig. 21. Gas injection plant in Lake Maracaibo.

April 12
President Leoni discusses with US President Johnson in Punta del Este, problems of mutual interest, among them the sulphur content of Venezuelan crudes and imports to the USA.

April 25
The US Department of Interior extends for 3 years the informal arrangement by which 4760 m^3/day of Mexican oil enters the US under the 'overland' exemption provisions of the mandatory oil import control programme.

In a speech to the First National Convention of Exporters, Minister Mayobre declares that he is optimistic regarding the future of the petroleum industry but that no one should expect a new boom, as in past years.

April 27
The CVP completes Mingo-1 as a new field discovery, south of Silvestre field.

May 4
Managing Director, Rubén Sáder Pérez, announces that the CVP

might run out of gasoline. Supply agreements had elapsed and oil companies were now asking for a 3-fold increase in prices.

May 10
The annual meeting of private investors, Fedecámaras, again sharply attacks national oil policy, following publication of some newspaper articles in the Caracas press.

May 12
President Leoni stresses national oil policy in a speech to the Nation. He says the Government would not accept any lowering of the reference prices and that the prospects of finding new reserves – in southern Lake Maracaibo, Gulf of Venezuela and the tar belt – are great.

May 13
The Economic Commission for Latin America ends its Twelfth Annual Meeting, held in Caracas. Reports on a seminar on oil economics for Latin America, held during February in Santiago, are discussed.

May 22
CVP completes Caipe-1 as a new field discovery. It is the northernmost producer in the Barinas basin.

May 23
President Leo Lowry of Creole informs the Senate Commission of Mines and Hydrocarbons that his company is not interested in signing service contracts with the CVP.

May 26
Creole submits to the Ministry of Mines and Hydrocarbons a proposal to build a US$100 million desulphurisation plant.

May 31
Former Development Minister M. Egaña, proposes to the National Economic Council that mixed companies should take over exploitation of concessions expiring in 1983.

Minister Mayobre defends national petroleum policy in a debate in the Senate.

1967

June 5
Israel–Arab war. Suez Canal is closed 2 days later; production in Arab exporting countries partially interrupted.

June 9
Thirteenth Conference of OPEC postponed.

June 12
Venezuelan officials declare country will not behave with 'mercantilist spirit' in Middle East crisis and that developments have shown 'geopolitical strategic value of Venezuelan oil'.

June 17
CVP makes first delivery of Boscán crude (11 900 m^3) to a private oil company, at Bajo Grande terminal.

June 25
The Ministry of Mines and Hydrocarbons authorises the building of 35 new service stations during 1967; 32 of the total 35 are given to the CVP (all of those in highly populated areas).

July 17
A Proclamation by the US President is issued to enhance the ability to provide adequate supplies of low sulphur residual fuel oil to use as fuel and to allow entrance into mandatory imports programme of new importers.

July 20
The National Congress passes a bill amending the 1943 Hydrocarbons Law. Service contracts may be negotiated by the Ministry of Mines and Hydrocarbons or by the CVP, on terms which should result more favourably for the nation than those of existing concessions. The duration of these contracts is 5 years for exploration and 20 years for development.

July 28
A law governing special agreements related to hydrocarbons desulphurisation is passed by the National Congress, to provide special reference prices for up to 10 years for low-sulphur petroleum crudes and products. Bill aims at allowing Venezuelan oil and products to meet more rigid requirements of US consumers.

July 31
Production during the month averages 600 034 m^3/day, an all-time monthly record.

August
Sinclair brings on stream a 900 m^3/day refinery in the Sinco field (*Barinas basin*) to meet local requirements of gasoline, kerosene and diesel oil.

Phillips and Shell put the injection plant Lamargás-1, in Lake Maracaibo, on stream, the first in the world to use gas turbines.

August 9
All concessions are now adapted to the 1943 Hydrocarbons Law.

August 14
An Executive Decree orders oil companies to cede to the CVP enough service stations to permit the national agency to cover 10% of the volume of gasoline sold in the domestic market during 1966.

August 18
Bases for regional common market of 'Andes countries' (Colombia, Chile, Ecuador, Peru and Venezuela) are agreed upon in Caracas.

August 28
An agreement for the integration of major fertilizer plants in Venezuela (El Tablazo) and Colombia (Barranquilla) is signed.

September 17
OPEC holds Extraordinary (Thirteenth) Conference in Rome.

September 26
Venezuela's Supreme Court hands down decision supporting the view of the Ministry of Mines and Hydrocarbons that concessionaire companies are obliged to submit all information the Ministry considers necessary to ensure full understanding of oil development in the country.

September 27
CVP announces that it will accept bids for the joint exploration and development in the South Lake Maracaibo offshore area, under service contracts.

CVP Managing Director Sáder-Pérez announces the start of production at 1300 tonnes/day from Mata (*Maturín basin*) to cover the volume sold to Petrobrás. It is the first important sale of the CVP in world markets and the Company is proud 'of having maintained prices'. Sáder-Pérez also announces that the supply of gasoline to the CVP stations is assured, as the agreement with Creole, Shell, Mene Grande, Texas, Mobil and Phillips is signed.

October 21
The Third General Assembly of ARPEL ends in Caracas. It is decided to move its headquarters to Montevideo and to study problems of economic integration within LAFTA, in the field of hydrocarbons.

November 1
Minister Mayobre declares Venezuela cannot always be dependent on what the USA will or will not do on oil imports. He also says that domestic political considerations are the major stumbling block for the elimination of discriminatory provisions in the US import programme, and that the proposed bills are an expression of a kind of economic isolationism.

November 3
The CVP asks interested foreign oil companies to participate in the sharing of information from a seismic survey carried out in the southern Lake Maracaibo area.

November 8
The Presidents of three large US gas transmission companies visit President Leoni and express a desire eventually to purchase, in liquid form, 20 million m^3 of natural gas per day.

November 9
Negotiations on a technical level resume between Venezuela and the USA referring to the new threat on oil trade.

November 10
The Second Meeting of Representatives of the National Oil Companies of OPEC ends in Djakarta. It is decided to authorise the commissioning of a study on the joint cooperation among the NOCs in the international market.

November 23
The National Congress unanimously backs the actions of the Ministry of Mines and Hydrocarbons in the USA against proposed restrictive legislation and urges Latin American Parliaments to take a unified stand on the matter.

November 29
The Fourteenth OPEC Conference ends in Vienna. It is decided to prepare new studies to make the production programme practical and effective and to admit Abu Dhabi as its ninth member.

December 1
The Chairman of US Federal Power Commission visits the Ministry of Mines and Hydrocarbons, probably to discuss the possibility of exporting liquefied natural gas to the USA.

December 4
A company for the distribution of natural gas is incorporated in Barquisimeto. The CVP and private interests participate.

December 19
The Senate transfer to the CVP 200 000 ha in the southern Lake Maracaibo area and 100 000 ha in the Paraguaná area.

Shell announces its readiness to negotiate desulphurising projects in Cardón.

December 26
The CVP announces exports of 160 000 m^3 of oil to Yugoslavia, commencing January 1968.

December 29
The CVP and Ecopetrol, the Colombian State oil agency, agree in Bogotá on non-overlapping geophysical surveys in the northern area of the Gulf of Venezuela.

December 31
The Government and Creole agree to build a desulphurising plant at Amuay. The unit valued at Bs528 million, will process 20 000 ton/day. It is the largest plant in the world.

The CVP signs an agreement to supply the new iron-ore plant of

Lagoven

Fig. 22. Terminal and refinery at Amuay.

Orinoco Mining with natural gas. The CVP will build a gas line from Mobil's Guario field to Guayana plant.

1968

January 1
The post of Councillor for Oil Affairs is established in the Madrid and Río de Janeiro Embassies.

US President Johnson limits investments outside the USA.

January 5
The Government and Creole sign an agreement to build a 16 000 m^3/day desulphurisation unit in Amuay, which will be the largest in the world.

January 9
The Fifteenth OPEC Conference (Extraordinary) ends in Beirut; discounts off posted prices are eliminated.

January 29
Proclamation 3823 of US President Johnson redefines tar sands and modifies the oil import programme, as a consequence of the Middle East war.

February 5
Minister Mayobre predicts that limits on US investments abroad will not affect the Venezuelan oil industry.

February 6
By Presidential Decree, State oil companies have the same rights and obligations as concessionaire companies.

March 1
The Petrochemical Institute is authorised to occupy the El Tablazo area.

March 18
The First Meeting of Experts of Arpel (Refineries) ends in Lima.

March 25
The first CVP crude oil cargo is shipped to Jugoslavia.

April 4
CVP and Philadelphia Gas Works agree to study the natural gas potential of free zones and national reserves in the Maturín basin.

May 2
The Ministry of Mines and Hydrocarbons and the Venezuelan Institute of Scientific Research agree to complete field work with a view to the exploitation of very heavy crude oil in the Orinoco bituminous belt.

May 3
The Government and Shell agree the construction of a desulphurising unit at Cardón refinery.

May 21
Minister Mayobre announces in Barquisimeto, that Canadian groups will be asked to participate in the building of a vanadium separation plant.

May 28
CVP and Industrialexport of Romania sign an agreement for the purchase of 11 million tonnes of crude oil and products over 10 years, in exchange for Romanian oil drilling equipment.

May 31
A meeting of the national oil companies of OPEC countries is convened in Vienna to revise a study on the co-ordination of international activities.

June 1
The Second Technical Petroleum workshop ends in Maracaibo.

June 12
Mito Juan acquires the Bella Vista oilfield.

June 25
The Sixteenth Conference of OPEC in Vienna agree on a Policy Statement, in relation to development, participation, reference prices, income tax stability, accounting, information, conservation and arbitration.

July 25
The Andean countries, except Venezuela, sign an accord on petrochemical tariffs.

September 10
Minister Mayobre pleads with the visiting Secretary General of OPEC on the need for an effective production apportionment programme.

October 9
The Technical Commission on Oil Statistics is established.

October 29
The third meeting of national oil companies of OPEC countries ends in Teheran.

November 6
The Fourth General Assembly of ARPEL ends in Santiago, Chile.

November 10
The Seventeenth Conference of OPEC ends in Baghdad, adopting regulations for the conservation of oil reservoirs, and a compromise is struck not to grant concessions to companies engaged in negative pricing practices.

November 14
The Central Bank assumes obligations to the International Monetary Fund.

November 28
First exports by Mito Juan, to Panama.

December 1
Rafael Caldera wins presidential elections.

1969

January 20
The Venezuelan Ambassador in Washington is confident President Richard Nixon will recognise the strategic importance of Venezuelan petroleum, as the programme of oil imports is revised.

February 1
The Ministry of Mines and Hydrocarbons declares that changes in the import policy of the USA, will not necessarily increase the level of residuals entering the Eastern Seaboard of the USA, although present export levels should be maintained.

February 11
Regulations on hydrocarbon reserves conservation are sanctioned.

February 23
US President Nixon announces a review of the oil imports programme.

1969

March 25
US President Nixon establishes a Cabinet Task Force on Oil Imports.

April 7
Home Stakes acquires Talon Petroleum.

April 11
The Society of Petroleum Engineers holds a public meeting on the ending of all oil concessions as of 1983.

April 20
The Minister of Mines and Hydrocarbons, Hugo Pérez-La Salvia, discusses the US oil imports programme in Washington, D.C.

April 30
A seminar entitled 'Venezuela – Year 2000' ends in Maracay. The general feeling is that oil might not be the basic economic activity in 2000, though still a very important one.

May 22
Venezuela quits a meeting of the Andean Pact in Cartagena, Colombia.

A 3-day meeting on the petroleum imports programme ends in Washington, D.C.

May 31
The Eighth Venezuelan Engineering Congress ends in Caracas, with a recommendation for the immediate creation of a committee to study all matters relating to the ending of oil concessions, commencing in 1983.

June 6
The Fourth Meeting of the Co-ordinating Commission of OPEC State Oil Companies ends in Benghazi, Lybia.

June 11
The National Congress agrees to back Government policy relating to the discrimination of Venezuelan oil by the USA.

June 19
The US Secretary of Interior orders the indefinite extension of the prohibition order placed on Phillips Petroleum not to import any oil products from outside the Western hemisphere for its Puerto Rico plant.

July 9
The Eighteenth Conference of OPEC ends in Vienna. Algeria is admitted as a new member, and a study is contracted to relate posted prices and reference prices with the prices of manufactured goods.

July 15
The National Front for the Defense of Venezuelan Petroleum requests from the Cabinet Task Force on Oil Imports the most favourable consideration, in the context of the general objectives of US petroleum policy. Creole Petroleum submits another study and a third is given to the Department of State by the Federal Government.

August 22
The Fourth Meeting of ARPEL experts (cost accounting) ends in Maracaibo.

November 7
A technical meeting between Venezuela and the USA on the oil imports programme ends in Caracas.

November 12
The National Front for the Defense of Venezuelan Petroleum submits to Washington a new document concerning the oil imports programme, on the premise that the Nixon Administration is sincere in its regard to Latin America.

November 21
The Fourth Venezuelan Geological Congress ends in Caracas.

December
CVP completes three strat holes for the Ministry of Mines and Hydrocarbons in the La Canoa area of the Orinoco bituminous belt.

1969

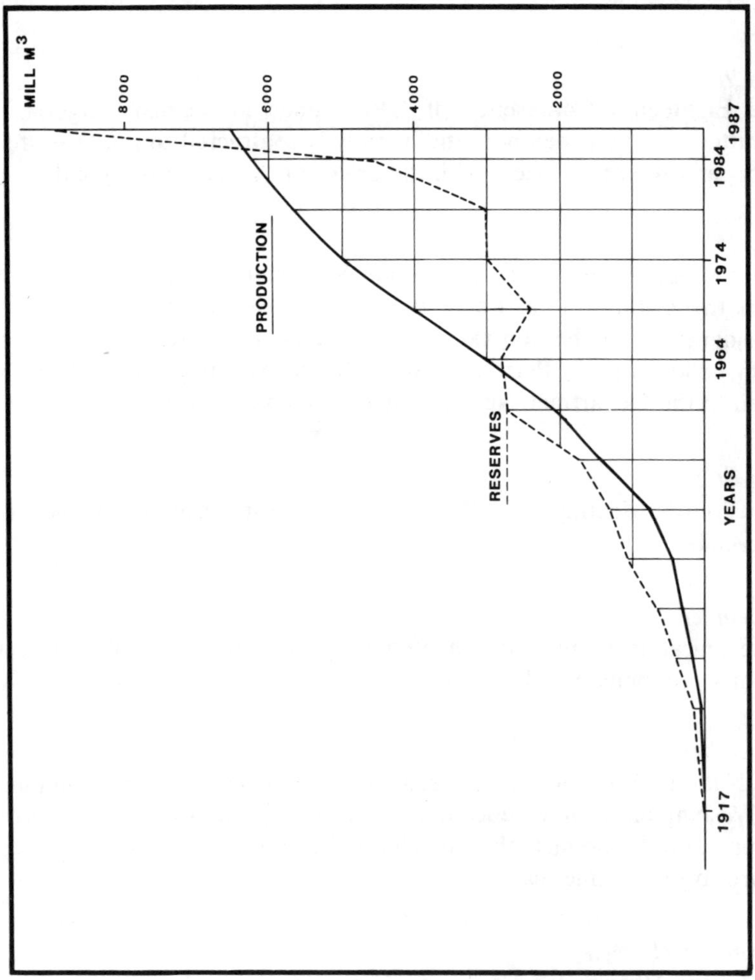

Fig. 23. Venezuela: production and reserves.

December 5
The Fifth ARPEL Assembly ends in Lima, Peru.

December 15
The first desulfurising plant in Venezuela goes on stream in Cardón refinery.

December 16
The Nineteenth OPEC Conference ends in Doha, Qatar.

December 24
The National Front for the Defense of Venezuelan Petroleum protests at the increased US discrimination of Venezuelan oil.

December 29
CVP and Philadelphia Gas Works sign an agreement for the export of 14 million m^3 of natural gas per day to the US Eastern Seaboard.

1970

January 14
The Supreme Court rules concessionaire companies are subject to municipal taxes.

February 6
A Meeting of the Interamerican Economic and Social Council ends in Caracas, agreeing to a joint committee for consultation and negotiation between Latin America and the USA.

February 9
The US Cabinet Task Force submits its report to US President Nixon.

February 17
The Asociación Pro-Venezuela and the Sciences Faculty of Caracas Central University sponsor a round-table discussion on the creation of a centre for petrochemical research.

February 20
US President Nixon rejects the Cabinet Task Force report. In Caracas, the National Front for the Defense of Venezuelan Petroleum reiter-

ates its request for the best possible consideration of Venezuelan oil, in the framework of a hemispheric policy.

March
Hideca, a Venezuelan marketing company, wins the tender for 400 000 tonnes of LPG to Gas del Estado (Argentina).

March 17
Asociación Pro-Venezuela asks the Research Council (CONICIT) to establish a group to study the creation of a petrochemical research centre.

March 18
In a telegram to the National Front for the Defense of Venezuelan Petroleum, the US State Department recognises the importance of their decision on the importation of crude oil and by-products.

April 2
CONICIT regard petroleum research as a high priority within national scientific and technological policy.

April 5
Mito Juan is the first Venezuelan oil company in 50 years to complete a new pool wildcat, Travieso-12.

May 20
A Bs200 million lawsuit, the largest in Venezuela, is filed against Shell in Maracaibo for evasion of municipal taxes.

June
Second edition of stratigraphic lexicon.

June 3
President Caldera defends Venezuelan petroleum policy in Washington, D.C.

June 17
The US adopts a small quota supplement of heating oil to hemisphere crudes.

June 26
The Twentieth OPEC Conference ends in Algiers, with the adoption of a production programme to safeguard prices, and the extension of the term of Secretary General to 3 years.

August 12
The idea advanced by the new Minister of Petroleum for Colombia, that any oil in the Gulf of Venezuela should be exploited jointly, was dismissed out of hand by other Latin American nations.

August 17
US President Nixon decides to continue the US quota system for imports of crude oil and its by-products.

September 8
The 228-km Anaco–Puerto Ordaz gasduct, with a capacity of 40 million m^3/day goes into operation.

September 16
The Minister of Mines and Hydrocarbons, Pérez-La Salvia, demands from concessionaire companies an explanation of why local prices do not follow increases in the international markets.

October 25
A meeting of the OPEC State Oil Companies ends in Kuwait, with an agreement to co-ordinate activities in world markets.

October 30
Creole sends its first shipment of elemental sulphur produced in the Amuay refinery to El Salvador.

November 18
The Minister of Mines and Hydrocarbons, Pérez-La Salvia, announces that sales will be halted if Venezuelan oil prices do not reflect any increases in the international price of oil. Opposition MPs had declared that the National Congress would intervene if concessionaires did not increase their oil prices.

December 3
Creole Petroleum and Shell deny having manipulated oil prices to reduce income tax obligations.

December 5
The Second Petroleum Congress ends in Caracas.

December 12
The Twenty-First OPEC Conference ends in Caraballeda, with the participants agreeing to increase income tax to at least 55%; to develop a system to protect oil earnings against the continued devaluation of natural resources and the revaluation of manufactured goods; and to demand Gulf oil prices be increased substantially.

December 17
The Income Tax Law is amended. The Federal Executive is empowered to fix the export values of Venezuelan crude oils and products for periods of 3 years, as the flat rate for income tax is set at 60%.

December 22
The discrimination against Venezuelan oil is accentuated when US President Nixon announces that Canada should receive half of the increase in the crude oil import quota.

1971

January 6
Regulations for the determination of export values are sanctioned.

January 15
In London, 13 major oil companies announce their willingness to negotiate with OPEC countries matters relating to the Caracas resolutions.

The US State Department sends a special Ambassador to the Middle East, in an effort to resolve the crisis emerging between producers and consumers.

February 4
The Twenty-Second OPEC (Extraordinary) Conference ends in Teheran, with a decision to increase prices unilaterally by 15 February and to impose an oil embargo if transnationals fail to accept these terms within a week.

February 14
A 5-year agreement is signed in Teheran between the Gulf States and the major oil companies, implementing Caracas Resolution XXI.120, whereby the crude oil price increases US$0.35 per barrel, automatic adjustments are to be implemented in 1973 and 1974, and there will be new density differentials.

March 8
The Ministries of Finance and Mines and Hydrocarbons announce that as of 18 March 1971, there will be a Bs16 per m^3 increase in export prices for income tax purposes, and that a freight supplement will be introduced as well as premiums for heavier oils.

March 9
CONICIT establishes a Petroleum and Petrochemicals Research Commission, co-ordinated by Aníbal R. Martínez.

March 25
President Caldera denounces the campaign in the USA directed at the unilateral setting of export prices by Government.

April
The Ministry of Mines and Hydrocarbons decides to designate as heavy, crude oil deposits in the Maturín basin, Orinoco Oil Belt.

May 5
Freight supplements on crude oil export prices are set.

June 4
US President Nixon invites Canada to form an energy common market.

June 19
In Caracas, the Director of Energy at the US State Department explains the proposal to form an energy common market with Canada.

July 1
The Municipal Council of Maracaibo creates an Energy Institute, which is to operate the domestic natural gas network.

1971

July 10
The Twenty-Third Extraordinary OPEC Conference in Vienna decides to suspend its production programme.

July 13
The Twenty-Fourth OPEC Conference in Vienna agrees Nigeria gain membership, and on a strategy for participation in concessions immediately.

July 21
Mito Juan and Petrom of Romania agree to complete 20 wells for CVP in the Boscán field.

Signal becomes Charter Venezuela Petroleum.

July 29
CVP signs service contracts for blocks A, D and E with Occidental Petroleum and for block B with Shell-Surca (*south Lake Maracaibo area*).

July 30
The Law on the Reversion of Concessions is sanctioned, whereby concessionaires relinquish all rights and permanent works without the payment of any indemnities whatsoever (Article 80 of the Hydrocarbons Law), in order to ensure continuity and efficiency of industry operations.

August 4
Those regulations in the Hydrocarbons Law relating to fiscal obligations derived from service contracts come into force.

August 26
The Law to Reserve to the State the Natural Gas Industry is sanctioned. Natural gas can only be exploited by CVP, with all concessionaires being obliged to donate all gas not used in their operations.

September 4
Mobil discovers a new field on completion of well Onado-51 (*Maturín basin, 30 km northeast of the Aguasay field*).

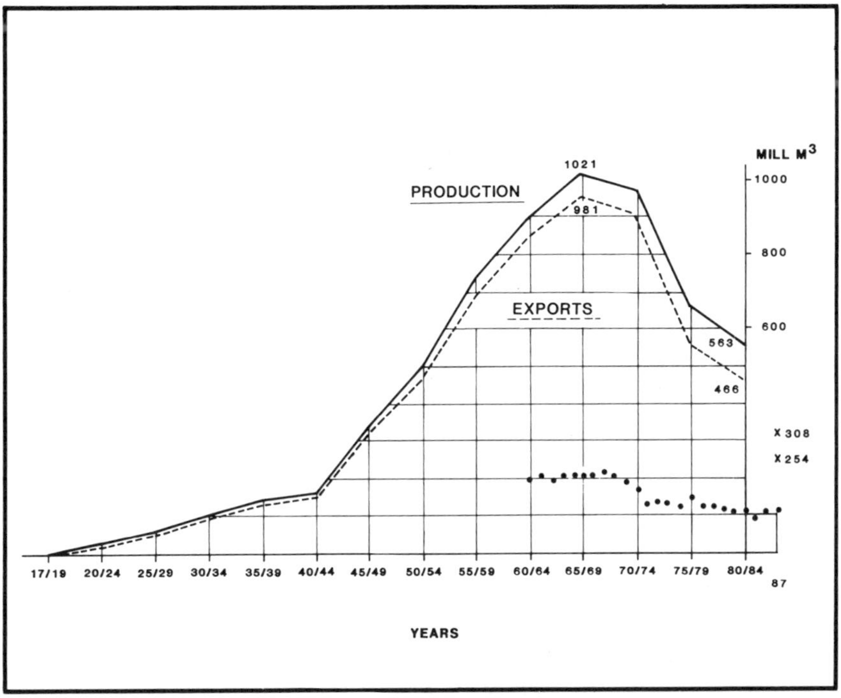

Fig. 24. Venezuela: production and exports.

September 16
CVP and Mobil sign a service contract for block C in the south Lake Maracaibo area.

September 22
In Beirut, the Twenty-Fifth OPEC (Extraordinary) Conference decides to adopt compensatory measures to alleviate the negative effects on earnings felt by Member countries due to the devaluation of the US dollar.

October
CVP buys a 50% share in the US marketing company Ven-Fuel.

October 14
The meeting of the managers of the central banks of OPEC countries ends in Vienna.

October 16
The Third Petroleum Workshop ends in Maracaibo.

December 5
The University of Zulia establishes an Oil Research Institute.

December 7
The Twenty-Sixth OPEC Conference ends in Abu Dhabi.

December 17
Decree 832 states that concessionaire companies are obliged to submit annual budgets for exploration, production, refining and marketing.

December 21
Export values for 1972 are set.

1972

January 1
President Caldera announces that the Commercial Reciprocity Agreement with the USA will end as of 1 July.

January 26
The first Regulations of the Reversion Law come into force.

February 12
Rómulo Betancourt sets the basic parameters for the nationalisation of the oil industry.

March 2
The Minister of Mines and Hydrocarbons, Pérez-La Salvia, informs concessionaires that the government will not accept reductions in production.

Juan Pablo Pérez-Alfonzo announces his alternative to both capitalism and socialism – worker-owned companies with both public and private investment.

March 7
The US Secretary of State says that Venezuela's importance as an oil supplier is in decline.

March 12
The Twenty-Fifth OPEC Conference ends in Beirut.

April 28
The Guarantee Fund as per the Reversion Law opens with Bs160 million deposits by concessionaire companies.

May
The Research Commission of CONICIT submits a proposal for the creation of INVEPET, the Petroleum Research Institute.

May 3
In Santiago, the Third UNCTAD Conference is told that during the last decade Venezuela subsidised US industry to the tune of US$8 billion.

June 9
The Twenty-Eighth OPEC (Extraordinary) conference, held in Beirut, supports Iraq's plans to nationalise its oil industry.

June 11
Japan shows an interest in offshore exploration and research in the Orinoco Oil Belt.

June 27
The Twenty-Ninth OPEC Conference ends in Vienna.

July 27
The US Ambassador in Caracas states that the USA is willing to exonerate Orinoco Oil Belt crudes of import tariffs.

August 7
CVP discovers the West Casma field on completion of well Acema-Casma No. 1 (*Maturín basin, 35 km northeast of the Junta field*).

August 10
A Division of Reversion is created within the Ministry of Mines and Hydrocarbons.

1972

August 21
The first meeting of the Petroleum Ministers of Latin American countries opens in Caracas.

September
The US Department of State proposes to both the Ministers of Foreign Affairs and Mines and Hydrocarbons, that the Orinoco Oil Belt be exploited by a joint private enterprise.

September 12
CVP tests 23 Miranda-2X, a second new field wildcat in the La Vela Embayment area, and finds 200 m^3/day of light oil (*25 km north of La Vela de Coro*).

October 24
The Minister of Mines and Hydrocarbons, Pérez-La Salvia, states that no further service contracts will be granted.

October 27
The Thirtieth OPEC (Extraordinary) Conference ends in Riyadh.

November 1
The National Front for the Defense of Venezuelan Petroleum proposes that the Orinoco Oil Belt should be declared a strategic reserve for national use.

November 2
The Second Consultative Meeting of the Energy Ministers of Latin America agrees in Lima to create OLADE (the Latin American Energy Organisation).

November 7
US President Nixon announces the Independence Project, aimed at making the USA self-sufficient in terms of energy.

November 16
The Ninth General Assembly of ARPEL and the Second Latin American Geological Congress end in Caracas.

November 30
The Thirty-First OPEC Conference ends in Lagos, Nigeria.

December
Venezuela and the USA discuss a protocol to determine the joint exploitation of the Orinoco Oil Belt.

December 13
The National Front for the Defense of Venezuelan Petroleum condemns the USA's suspension of the hemispheric preference on its small volume of fuel oil imports.

December 14
On the 50th anniversary of the blow-out of Los Barrosos No. 2, Fernando Chumaceiro (Corpozulia), Diego Arria (Corpoturismo) and Aníbal R. Martínez (National Front for the Defense of Venezuelan Petroleum) sign the incorporation of the Petroleum Museum S.A. in Cabimas.

December 20
Foninves is created by Law, a fund for hydrocarbons research and personnel.

1973

Phillips begins a secondary recovery project using continuous hot water injection in the Morichal field.

January
There is an 'energy crisis' in the USA.

January 18
US President Nixon decides to increase import quotas to 65% and to suspend controls on purchases of heating oils.

January 28
The Minister of Mines and Hydrocarbons, Pérez-La Salvia, declares that the Orinoco Oil Belt will not be negotiated and that only CVP will participate in future operations.

February 5
The USA promotes the creation of the International Energy Agency, to counter OPEC actions.

1973

March 9
The Director of the US Office of Energy Preparedness indicates his Government's interest in exploiting the Orinoco Oil Belt jointly with Venezuela.

March 17
The Thirty-Second OPEC Conference ends in Vienna.

March 23
US President Nixon cancels oil import restrictions to ease the 'energy crisis'.

April 3
CVP completes 23 Miranda-4X as a new field discovery well (*La Vela Embayment basin, 11 km north of La Vela de Coro*).

April 9
The Pro-Venezuela Association asks the Government for a clear, definitive policy on the Orinoco Oil Belt.

April 18
US President Nixon ends the compulsory oil import programme and begins a system of quotas.

May
Oil products not available in some international markets.

May 24
Officials of the European Common Market show an interest in the joint development of the Orinoco Oil Belt.

May 26
The Thirty-Third OPEC Conference ends in Vienna.

June 2
Following the Geneva meeting with OPEC countries, transnational oil companies agree to increase prices by 12%.

June 12
Occidental completes discovery well SLA-6-2X in the service contract area of south Lake Maracaibo (*20 km south of the Lamar field*).

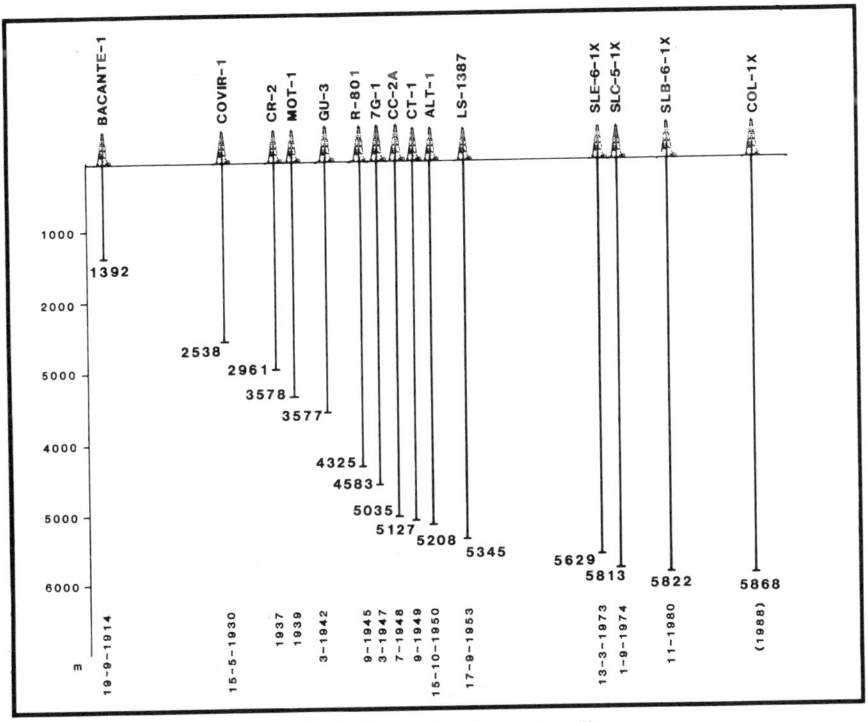

Fig. 25. Venezuela: deepest wells.

June 15
CVP completes 23 Miranda-5X as a new field (*La Vela Embayment basin, 15 km north of La Vela de Coro*).

June 21
A Law is sanctioned to reserve internal markets to the State.

June 28
The Thirty-Fourth OPEC Conference in Vienna accepts Ecuador as an Associate Member and agrees on a policy statement with a view to attain greater access to technology and to strengthen cooperation with oil-importing nations in the developing world.

August
CVP completes a programme of 14 strat wells in the Orinoco Oil Belt for the Ministry of Mines and Hydrocarbons.

August 22
The First International Symposium of Vanadium in Petroleum ends in Maracaibo.

September 16
A Ministerial Committee of OPEC Members is appointed by the Thirty-Fifth OPEC Conference in Vienna to revise the terms of the Teheran, Tripoli and Lagos agreements with oil companies.

September 28
Fifth adjustment on prices during the year – up 17% for heavy oils and 22% for light oils.

October 22
Occidental completes SLE-4-2X as discovery of a new condensate field, at 5395–5496 m, it is the deepest producer in Venezuela (*service contract area, Maracaibo basin, 50 km south of the Lamar field*).

October 26
A joint Resolution by the Ministries of Finance and Mines and Hydrocarbons increase export prices by 50%.

November 2
The Latin America Energy Organisation (OLADE) is formally created in Lima, with its headquarters to be in Quito.

November 20
The Thirty-Sixth OPEC Conference in Vienna admits Ecuador as a full Member and Gabon as an associate Member.

November 28
The National Commission for Nuclear Matters is created.

November 30
Export values are set 82% higher than at the beginning of the year.

December 9
Carlos Andrés Pérez is elected President.

December 17
The first stage of the El Tablazo petrochemical complex goes into operation.

1974

January
A study by the Ministry of Mines and Hydrocarbons is completed on 95 'marginal' fields. There are 330 service companies registered. Concessionaire investments total Bs21 billion.

January 2
Creole is asked to return to the State, without compensation, the Mulata and Jusepín areas of the Jusepín field.

January 9
The Thirty-Seventh OPEC (Extraordinary) Conference approves the request of Abu Dhabi to transfer membership to the United Arab Emirates.

January 28
The Ministry of Mines and Hydrocarbons starts a Well Data Bank.

February 4
Creole hands Mulata and Jusepín areas over to CVP.

March 17
The Thirty-Eighth OPEC Conference ends in Vienna.

March 22
Decree 10 of President Pérez establishes a Commission to study the end of concessions.

April
First oil production drop.

April 7
The OPEC Development Fund is created.

May
Venezuela contributes US$540 million to an International Monetary Fund for developing countries.

1974

May 16
President Pérez announces the creation of an oil holding company and compensation to concessionaires on the basis of non-depreciated assets.

June
The export values of oil increase by 15%.

June 1
The Ministry of Mines and Hydrocarbons makes correction of oil density by water content compulsory.

June 15
The Third Petroleum Congress ends in Caracas.

June 17
The Fortieth OPEC Conference ends in Quito.

June 19
The Petrolia site in the La Alquitrana field is made a National Monument.

July 6
CVP suspends the 23 Miranda-12 as discovery well of a condensate and natural gasfield (*La Vela Embayment basin, 8 km northeast of La Vela de Coro*).

July 11
The Venezuelan Investment Fund comes into operation, made up of extraordinary oil earnings.

August 7
CVP resumes strat drilling in the Orinoco Oil Belt for the Ministry of Mines and Hydrocarbons.

August 16
The Ministry of Mines and Hydrocarbons appoints observers to each oil company.

August 29
The first meeting of the Third UN Conference of the Sea ends in Caracas.

September 3
The National Coal Board is created.

September 18
Exploration of the La Vela Embayment basin is suspended by CVP.

October 6
Exploration of the service contract area in south Lake Maracaibo is completed.

CVP and Shell sign an agreement to investigate the demetalisation of heavy Venezuelan crude oil.

November 7
CVP spuds two wells in the Orinoco Oil Belt for productivity tests.

November 11
The First Petroleum and Mining Congress ends in Caracas.

November 15
CVP discovers a new field on completion of the exploratory well Casma-1 (*Maturín basin, 30 km east of the Junta field*).

November 16
The First National Workshop on Refining ends in Punto Fijo.

December 13
The Forty-Second OPEC Conference ends in Vienna.

December 23
The Presidential Commission submits a report proposing the organisation of an oil administration for the national industry.

December 31
Total petroleum investments in Venezuela stand at Bs30 762 million of which Bs15 454 million is depreciated.

Proved reserves increase to 2952 million m^3, as a result of a new evaluation by the Ministry of Mines and Hydrocarbons.

INTEVEP

Fig. 26. INTEVEP, the research institute of Petróleos de Venezuela, has developed relatively quickly into a major centre.

1975

January

The Government opens up debate and consultation in relation to nationalisation, on the basis of a draft bill by the Reversion Commission.

The second stage of strat drilling in the Orinoco Oil Belt ends.

January 10

An agreement is signed for the possible construction of a proteins from hydrocarbons plant by the Ministry of Mines and Hydrocarbons, British Petroleum and the Association of Makers of Concentrates for Animals.

January 14

The National Council for the Petrochemical Industry is created.

February
CVP ends 66 strat holes in the Orinoco Oil Belt.

February 27
The Forty-Third OPEC Conference ends in Vienna.
The Fifth Ministerial Meeting of OLADE concludes in Kingston, Jamaica.

March 3
Creole cedes operation of the Cumarebo field to CVP.

March 6
Solemn Declaration of Sovereigns and Heads of State of OPEC countries in Algiers, on world economic situation and international cooperation.

April 9
The Socialist MEP party denounces the Government for secretly negotiating nationalisation with concessionaire companies.

April 22
INVEPET finishes a report on technology transfer in the oil industry.

April 29
A Permanent Commission and Secretariat is created within the National Energy Council.

May 7
In Caracas, President Pérez and the Shah of Iran decide to organise a joint oil tanker company.

May 21
The Public Accountants College denounces the 'occult liabilities of Bs5 billion in the books of concessionaire companies in favour of the nation.

June
Bioproteínas de Venezuela is organised as a joint venture between the Government and BP.

June 11
The Forty-fourth OPEC Conference ends in Libreville, Gabon, deciding to use DEGs as the accounting monetary unit for payments. Gabon is accepted as the thirteenth Member Country.

June 24
Exxon acquires all Creole Petroleum shares.

July 5
On Venezuela's Independence Day, President Pérez announces that the petroleum industry holding company started recruiting personnel.

July 12
The First National Congress of Science and Technology ends in Caracas.

August 26
The Ministry of Mines and Hydrocarbons authorises new tariffs for domestic natural gas served in Maracaibo.

August 29
The Organic Law to reserve to the State the industry and commerce of hydrocarbons is sanctioned.

August 30
By Presidential Decree 1123, the holding company of the petroleum industry, Petróleos de Venezuela, is created, to plan, coordinate and supervise all activities. The President of the Board is Rafael Alfonso Ravard.

September
Petróleos decides that CVP shall no longer monopolise the domestic market.

September 2
Executive Decree 1127 converts CVP into a mercantile society.

September 10
The Supervisory Commission of Petroleum Installations is sworn in. José Martorano Battisti is Co-ordinator.

September 11
Oil Minister Hernández-Acosta announces that the national companies succeeding each concessionaire company shall be Lagoven (Creole Petroleum), Maraven (Shell de Venezuela), Meneven (Mene Grande), Deltaven (Texaco), Palmaven (Sun), Amoven (Amoco), Boscanven (Chevron), Vistaven (Mito Juan), Llanoven (Mobil), Roqueven (Phillips), Guariven (Las Mercedes), Bariven (Sinclair and Atlantic) and Taloven (Talon).

September 12
The Sixth Ministerial meeting of OLADE ends in Mexico City.

September 15
Petróleos de Venezuela is registered as a mercantile company, with all of its shares owned by the Republic of Venezuela.

September 20
The First Workshop of Geology, Mines and Oil ends in Maracaibo.

With one negative vote, the CVP Board agrees that Occidental's discovery in block E of the service contracts is commercial.

October 1
Export values are set, as recommended by OPEC, and income tax is increased to 72%.

October 10
President Pérez orders that payments to Occidental should be stopped, until the outcome of an investigation into the bribing of Government officials is completed.

October 13
Concessionaires receive offers of indemnification.

October 29
Indemnification offers totalling Bs4317 million are accepted by the concessionaires, of which Bs2400 million is for a Guarantee Fund.

November
The M-6 continuous steam injection project starts at Tía Juana (*Bolívar Coastal field*).

December
There is a public debate regarding indemnification and other payments to concessionaires.

December 2
CVP is converted to a mercantile company, with all of its shares owned by Petróleos de Venezuela.

December 15
In preparation for the cancellation of concessions, production drops by half.

December 17
Congress approves all expropriation agreements with concessionaires, including indemnification.

December 21
The Headquarters of OPEC in Vienna are taken over by a guerilla group during the Forty-Sixth Conference. Kidnapped Ministers are released in Algiers 1 week later.

December 31
All concessions are cancelled. The Ministry of Mines and Hydrocarbons determines areas of responsibility for succeeding national companies. The Guarantee Fund of Bs2 357 million is constituted.

1976

January 1
At the Zumaque-1 site (*Mene Grande field*), President Pérez declares that the industry and commerce of hydrocarbons be reserved to the State.

Petróleos de Venezuela effectively assumes responsibility for all industry operations, acquiring the shares of 14 national companies – which succeeded the concessionaires – at par value. The Research Institute, Intevep, is also acquired by Petróleos.

January 2
Petróleos de Venezuela sign technology support and marketing agreements with each of the former concessionaires. A royalty shall be paid on the basis of production and refining volumes.

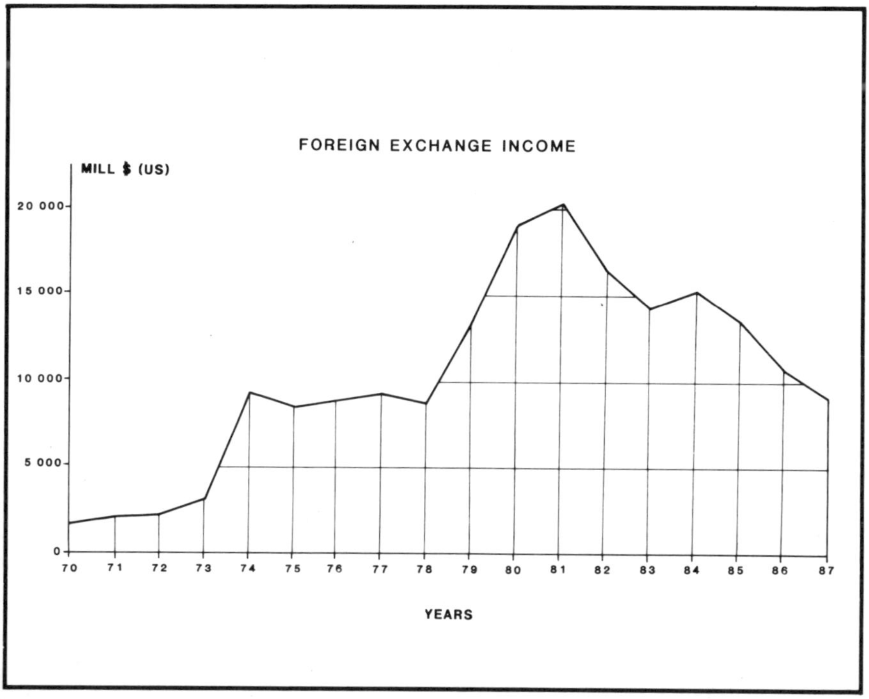

Fig. 27. Petróleos de Venezuela: foreign exchange income.

January 26
The Training Institute of the national industry commences operation.

January 28
The Finance Ministers of OPEC Member Countries create the OPEC Fund.

March
A study of a new Hydrocarbons Law begins.

March 3
A contract is signed with six Japanese companies to handle up to 5 millon ton/year.

March 4
FOB prices of Venezuelan crude oils are set for the year.

April 15
The National Comptroller, José Muci-Abraham, serves former concessionaires with claims of Bs2300 million on income tax returns.

April 27
The Eleventh ARPEL Assembly meets in Santiago.

May
Atlantic withdraws from Venezuela, as they find the technology support agreement unacceptable.

May 6
Lagoven sell 320 000 m^3 of Bachaquero crude oil to the Ivory Coast.

May 7
Petróleos de Venezuela and Empresa Española de Petróleos agree to the sale of 3000 m^3/day of medium and heavy crude oils.

May 10
Regulations are issued for the physical certification of assets of former concessionaires.

May 11
The OPEC Fund agrees in Paris to contribute US$400 million to the Agricultural Development Fund.

May 28
The Forty-Seventh Conference of OPEC ends in Bali, Indonesia.

June 5
The Congress investigating committee condemns Occidental on charges of fraud and illicit operations.

July 1
The dollar value is set at Bs4·28.

August 20
Income tax is reduced to 65·6%.

September 29
Venezuela and the USSR agree on exchange visits of petroleum engineers.

November
An aeromagnetometer survey of the Carrizal area (*Orinoco Belt field*) is completed.

Petróleos de Venezuela (UK) and Petróleos de Venezuela (USA) are incorporated in London and New York, respectively.

November 19
President Pérez signs a bilateral agreement on the transfer of oil technology in Rome.

November 20
The First National Geophysical Convention ends in Ciudad Bolívar.

The Second Technical Workshop on Refining ends in Puerto La Cruz.

November 26
In Moscow, President Pérez signs an agreement to exchange crude oil delivered to Cuba with Black Sea crude oil delivered to Petróleos' clients in Europe.

December 1
The operating subsidiaries of Petróleos de Venezuela are reorganised. Amoven is merged with Lagoven, Roqueven with Maraven, Boscanven with CVP and Bariven, and Taloven and Vistaven with Meneven.

December 2
First collective contract of Petróleos de Venezuela and oil workers' unions.

December 17
The Forty-Eighth conference of OPEC ends in Doha, Qatar.

1977

February 3
Shipment of 182 000 tonnes of crude oil out of Puerto Miranda.

1977

March 25
Mitsubishi buy 250 000 tonnes of crude oil.

April 1
The Ministry of Mines and Hydrocarbons becomes the Ministry of Energy and Mines.

April 27
The Twelfth Assembly of ARPEL ends in Lima.

May
Las Mercedes receives an exploration permit in Suriname.

May 22
A group of university professors and politicians, in an open letter, denounce the inconvenience of the secret technology and marketing agreements signed by Petróleos de Venezuela.

June
CVP gives back all service stations received during the last 12 years as per Decree 187.

June 25
Lagoven completes UD–130 as first producer from Cretaceous Río Negro formation (*Urdaneta field*).

June 29
The Socialist MEP party asks the Supreme Court to annul the technological agreements.

July
Petróleos de Venezuela and British Petroleum postpone the bioproteins project indefinitely.

July 2
The Fifth Technical Petroleum Workshop ends in Caracas.

July 13
The Forty-Ninth Conference of OPEC ends in Stockholm.

July 18
The Petrochemical Institute is converted into a mercantile society.

September 8
The Seventh Ministerial Meeting of OLADE ends in Quito.

October
Responsibility for the evaluation of the Orinoco Belt field is given to Petróleos de Venezuela by the Federal Executive.

October 12
The Second Seminar of OPEC on the present and future of national oil companies ends in Vienna.

November 22
Presidential Decree 2454 makes the Petrochemical Institute a mercantile society.

November 23
The Fifth Venezuelan Geological Congress ends in Caracas.

December 21
The Fiftieth OPEC Conference ends in Caracas.

December 27
The first liquid sulphur cargo – 15 000 tonnes – is shipped from the Amuay refinery to Tampa, Florida.

1978

January
The Ministry of Energy and Mines considers a resumption of the exploitation of the Guanoco asphalt lake.

February 16
The Government and former concessionaires agree on the total amount to be paid to the Guarantee Fund, in relation to real assets.

March 1
The Federal Executive transfers the ownership of all of the Petrochemical Institute shares to Petróleos de Venezuela.

1978

March 28
In Caracas, Venezuela and the USA sign a treaty on the delimination of marine areas in the Caribbean sea, in consideration of the Las Aves and Los Roques archipelagos and La Orchila and Aves islands facing Puerto Rico and St Croix islands, respectively.

March 29
The Ministry of Energy and Mines authorises Petróleos de Venezuela to explore in any part of the national territory.

March 31
Venezuela and The Netherlands sign a treaty on the delimination of marine areas in the Caribbean sea in Willemstad, Curaçao. Article 6 makes compulsory unification of any petroleum deposits astride the binational boundary.

April 7
The First OLADE Seminar on Energy Planning ends in Caraballeda.

May
The Society of Petroleum Engineers objects to the construction of a flexicoker unit in the Amuay refinery.

June
The Geology Directorate of the Ministry of Energy and Mines publishes the geological–structural map of Venezuela.

June 19
The Fifty-First OPEC Conference ends in Geneva.

July
INTEVEP moves to its new headquarters in El Tambor, near Los Teques, 25 km southwest of Caracas.

August
Petróleos de Venezuela's responsibility for the exploration of the Orinoco Belt field is shared among the operating subsidiaries Lagoven, Meneven, Maraven and CVP-Llanoven, creating a public outcry.

The consumption of motor gasolines is over 20 million litres/day.

August 4
The First Seminar on the Technology of Refinery Asphalt ends in Caracas.

August 23
The National Congress sanctions a Law in agreement with the international convention on sea pollution control.

September
The dyke protecting the Tía Juana area in Lake Maracaibo, is raised 2 m as subsidence continues.

September 3
The first centenary of the Venezuelan petroleum industry.

September 6
The Ninth Ministerial Meeting of OLADE ends in Mexico City.

September 22
The First Venezuelan Energy Congress ends in Caracas.

October 12
Lagoven spuds in Guarao-1, the first exploration well offshore in the Atlantic ocean (*Maturín basin, 75 km northeast of the Boca Grande, Orinoco River delta*).

October 16
Venezuela and West Germany sign an agreement on technical cooperation and research in the hydrocarbons sector.

October 29
Maraven spuds in GT-1X, the first well in the Triste Gulf, offshore the Falcón basin.

November
Llanoven merges with CVP – the new company is named Corpoven.

November 3
A seminar on exploration and operations techniques in the Cretaceous ends in Caracas.

December
A lateral radar survey is completed north of the 8°N parallel.

December 17
The Fifty-Second OPEC Conference ends in Abu Dhabi.

December 31
Heavy crude oil and residuals account for 25% of exports; sales to non-traditional clients accounts for 37% of exports. Natural gas production is down 2%, whereas motor gasoline production is up 24%.

1979

March 3
Venezuela and the Dominican Republic sign a treaty on the delimitation of marine areas in the Caribbean sea.

March 5
The Ministry of Energy and Mines publishes its first document on national energy policy.

March 12
Luis Herrera-Campins takes office as the new President of the Republic. Humberto Calderón-Berti becomes Minister of Energy and Mines.

March 27
The Fifty-Third OPEC (Extraordinary) Conference ends in Geneva.

April 27
On its tenth anniversary, the National Front for the Defense of Venezuelan Petroleum becomes a non-profit civil association.

May 31
INTEVEP becomes a mercantile society, with all of its shares being owned by Petróleos de Venezuela.

June 1
The Fourteenth Assembly of ARPEL ends in Río de Janeiro.

June 8
Venezuela and Canada sign a technological cooperation agreement for research on the exploitation and processing of very heavy crude oil.

June 25
Patao-1 is completed by Lagoven as the first discovery well of a giant natural gasfield in the Margarita basin (*40 km north and offshore of the Paria Cape, 130 km southeast of the Los Testigos archipelago*).

June 28
The Fifty-Ninth OPEC Conference ends in Geneva.

July
The Ministry of Foreign Relations returns, unopened, a letter of the US Ambassador, regarding a law suit filed against OPEC in a Californian court, thus reaffirming Venezuela's stance on a nation's sovereignty over its natural resources.

The debate continues on the immediate development of the Orinoco Oil Belt.

July 7
The First Extraordinary Meeting of the Ministers of OLADE ends in San José, Costa Rica.

July 11
Venezuela and the USSR agree to complete a drilling programme to test down-hole turbine drilling equipment.

August
A revision of technology support contracts commences.

August 23
The number of Board members of Petróleos de Venezuela is increased to nine, and their term of office is reduced to 2 years.

September
Lagoven begins to develop the Cerro Negro principal producing area and Meneven the Hamaca principal producing area (*Orinoco Belt oilfield*).

September 30
Venezuela has produced one tonne of every 5 tonnes of crude oil used in the world, since the inception of the industry 120 years earlier.

October 30
In the Petrofina law suit, filed by the General Comptroller regarding the payment of income tax, the Supreme Court rules in favour of the Nation.

November
In London, Maraven and Shell discuss a new basis for technology support agreements.

The olefins plant in the El Tablazo petrochemical complex starts operation.

November 1
Lagoven discovers a Cretaceous limestone field on completion of San José-1 (*Maracaibo basin, 110 km southwest of Maracaibo, 35 km north of the Alturitas field*).

November 2
The Sixth Technical Petroleum Workshop ends in Maracaibo.

November 15
The Ministry of Energy and Mines determines the geographical areas of operation for each subsidiary of Petróleos de Venezuela.

December
Pequiven denies causing irreparable damage to the sea and rivers near its Morón plants, due to mercury poisoning.

December 4
Venezuela and Ecuador sign a treaty for cooperation in oil matters.

December 11
The Tenth Ministerial Meeting of OLADE ends in Panamá City.

December 20
The Fifty-Fifth OPEC Conference ends in Caracas.

Lagoven

Fig. 28. En route to the drilling of a new field wildcat in Margarita basin.

December 31
Aid is extended to Central American and Caribbean countries as crude oil sales for the year amount to US$160 million.

End of technological support agreements.

1980

Foninves publishes *Geology of Venezuela and its Petroliferous Basins* by Clemente González de Juana and others.

March
More than 100 marketing agreements are signed by Petróleos de Venezuela and its affiliates with former concessionaires.

INTEVEP and the French Petroleum Institute sign an agreement for joint research and cooperation on petroleum matters.

March 3
Venezuela gives technical assistance relating to oil to Grenada and St. Lucia.

1980

March 23
Lagoven discovers a giant natural gasfield on completion of Mejillones-1 (*Margarita basin, 20 km west of the Patao field*).

April
Subsidiaries of Petróleos de Venezuela sign much revised technological support agreements with Exxon, Gulf and Shell, and new agreements with British Petroleum and Compagnie Française du Pétrole.

April 22
The Fifth OPEC State Oil Companies Meeting ends in Vienna.

April 24
The Minister of Energy and Mines, Humberto Calderón-Berti, announces that Venezuela shall not sell oil to the USA for the strategic reserve.

May
Lagoven commences an experimental continuous steam injection project in the Jobo field.
Schlumberger sponsors a meeting on formation evaluation.

May 8
The Fifty-Fifth OPEC (Extraordinary) Conference ends in Taif, Saudi Arabia.

May 11
Lagoven discovers a new oilfield on completion of SOL -2 (*Maracaibo basin, 60 km southwest of the Lagofield*).

May 20
Pure Oil accepts to pay in full, the General Comptroller's income tax claim.

May 22
The National Council for the Development of Capital Goods Industry is created.

May 29
The Fifteenth Assembly of ARPEL ends in Buenos Aires.

June 4
Ministry of Energy and Mines officials and Petróleos de Venezuela managers meet in Puerto La Cruz to evaluate the state of the industry and improve communications.

June 11
The Fifty-Seventh OPEC Conference ends in Algiers, with recommendations to create a news agency and to increase the marker price to US$32/barrel.

June 27
A collective contract is signed by Petróleos de Venezuela and the Oil Workers' Unions' Federation.

June 30
The Ministry of Energy and Mines set the prices of wax in inland market.

July
The Norwegian company Geco finishes a seismic survey of the Aves and Paria platform (*Margarita basin*).

The Petróleos de Venezuela subsidiary Bariven is given the responsibility for the acquisition of materials and equipment outside Venezuela.

July 10
Petróleos de Venezuela and the Ministry of Environment sign an agreement to formulate a national plan of development of petroleum areas.

July 17
Venezuela and France agree to the delimitation of marine and submarine areas in the Caribbean between Aves island and Guadeloupe and Martinique islands.

July 18
Venezuela and the USA agree to a joint investigation of the physicochemical characterisation and enhanced recovery of heavy crude oils.

August
The first research agreement is signed by INTEVEP with a local university (Central).

August 3
Venezuela and Mexico sign the San José agreement to supply, on advantageous terms, more than 17 000 tonnes of crude oil per day to several countries in Central America and the Caribbean.

August 28
Lagoven discovers a giant natural gasfield on completion of Dragón-1 (*Margarita basin, 30 km east-northeast of the Patao field*).

Petróleos de Venezuela and Veba Oel of West Germany agree to a technical and commercial cooperation programme.

September
Venezuela and the USA renew their bilateral energy cooperation treaty.

September 12
An informal Ministerial Meeting of OLADE ends in Caracas.

September 17
The Fifty-Eighth OPEC (Extraordinary) Conference ends in Vienna.

A meeting of the Finance and Foreign Relations Ministers of OPEC Member Countries ends in Vienna.

September 27
Lagoven discovers a giant natural gasfield on completion of exploration well Los Testigos-1 (*Margarita basin, 95 km northwest of the Patao field and 40 km east-northeast of the Los Testigos archipelago*).

October 3
INTEVEP seminar on the processing of heavy crude oil with participants from Canada, USA and France, ends in Caracas.

October 20
An agreement between Venezuela and Colombia for the partition of marine areas in Gulf of Venezuela is drafted in Caraballeda, including a proposal for the equal ownership of any common petroleum reservoirs, later repealed.

October 30
Lagoven discovers a heavy crude oilfield in the Paria gulf, on

Fig. 29. Orinoco Belt field.

completion of Morro-1 (*Maturín basin, 35 km northeast of the Pedernales field*).

November 3
The news agency OPECNA starts operations.

December 16
The Fifty-Ninth OPEC Conference ends in Bali, Indonesia, fixing the marker price at US$32/barrel.

December 31
Foreign liquid assets of Petróleos de Venezuela surpasses US$10 billion.

1981

INTEVEP starts research on the emulsion of Orinoco natural bitumen in water.

January
First issue of Intevep *Revista Técnica.*

January 1
The Ministry for Energy and Mines increases oil prices by 10%, thus following OPEC recommendations.

March 17
The Jobo field experimental continuous steam injection project starts.

March 21
An inter-institutional commission is created to co-ordinate land development in the Orinoco Oil Belt area.

May 8
In Cabimas, President Herrera-Campíns signs an inter-institutional agreement to initiate a global organisational programme for the eastern coast of Lake Maracaibo.

May 26
The Sixtieth OPEC Conference ends in Geneva with recommendations to reduce production and maintain prices.

June
Oil production in OPEC countries is reduced by 50%, thus complying with the Conference resolution.

July 29
The Sixteenth Assembly of ARPEL ends in Quito.

August 3
Venezuela and Mexico extend the San Jośe agreement.

August 26
Meneven's headquarters are moved to Puerto La Cruz and Pequiven's headquarters to Maracaibo.

September
Minister Calderón accuses Texaco of refusing to process Venezuelan crude oils in Managua refinery.

October 6
President Herrera announces that Venezuela shall be the world centre of heavy and very heavy crude oil research and development.

October 16
In Caracas, Venezuela, Brazil and Mexico agree to form an international oil company.

October 29
The Sixty-First OPEC Conference ends in Geneva.

November
Elf-Aquitaine concludes a study of the possible development of giant natural gasfields, discovered in the Margarita basin, for Lagoven.

November 1
Venezuela and China sign an agreement for science and technological cooperation.

November 8
The Twelfth Ministerial Meeting of OLADE ends in Santo Domingo, with the approval of a Latin American Energy Cooperation Programme.

December 10
The Sixty-Second OPEC Conference ends in Abu Dhabi, with an agreement to reduce the reference price of some crudes and to adjust differentials among Member Countries.

The Institute for Control and Conservation of Lake Maracaibo is created.

December 19
Lagoven discovers its first condensate field in the Margarita basin, on completion of Río Caribe-1 (*24 km north of the Paria peninsula, 50 km west of the Patao field*).

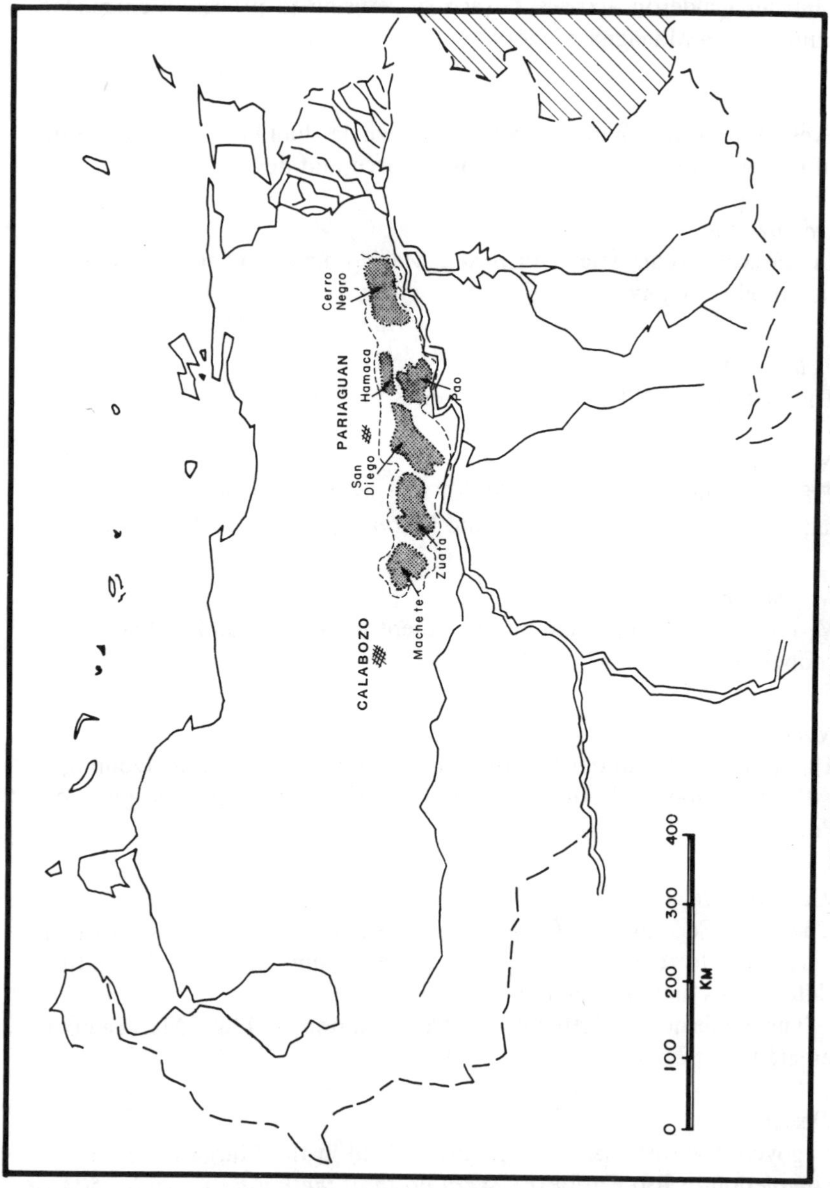

Fig. 30. Orinoco Belt field: main producing areas.

December 23
Pequiven cedes the distribution and marketing of fertilisers in the domestic market to Palmaven.

1982

January
Petróleos de Venezuela announces the *in situ* volume of the Orinoco Belt field to be 160 billion m^3.

February
The impact of the sharp drop in oil prices on fiscal income and national programmes is debated.

February 1
Lagoven discovers natural gasfield South Patao (*Margarita basin, 15 km south of the Patao field*).

February 17
The Second International Conference on Heavy Crude and Tar Sands ends in Caracas.

April
Venezuela withdraws some US$700 million from British banks, due to the Malvinas (Falklands) Islands conflict in the South Atlantic.

April 7
Emergency economic measures, with an immediate reduction of US$2 billion in public expenditure. High octane gasoline prices rise by 200%.

May
Minister Calderón-Berti refuses information on Venezuelan supplies of oil products to Argentina in the South Atlantic war.

Production is reduced to 235 000 m^3/day to comply with OPEC directives.

May 7
The Seventeenth ARPEL Assembly ends in Cartagena, Colombia.

1982

May 26
Corpoven ends its refinery expansion in El Palito.

July 2
The Third Latin American Oil Exhibition closes in Caracas.

August 4
Petróleos de Venezuela and Veba Oel of West Germany sign a preliminary commercial and technological accord.

September
Texaco agrees to the processing of Venezuelan crude oils in its Managua refinery.

September 27
The National Executive decides to centralise the administration of foreign currencies of Petróleos de Venezuela, in the Central Bank whereby the bolívar current accounts in the Bank would be credited with the equivalent in foreign currencies.

September 30
The San Lorenzo refinery is shut down after 65 years of operation.

October 16
Lagoven discovers a natural gasfield on completion of Caracolito-1 (*Margarita basin, 60 km northwest of the Patao field and 30 km southeast of the Los Testigos field*).

October 26
The College of Engineers opposes the creation of joint-venture companies to exploit the Orinoco Oil Belt or to participate in certain operations.

October 28
An agreement is signed by Petróleos de Venezuela and the Central Bank on the use and transfer of foreign currencies.

November 25
The flexicoker unit at the Amuay refinery – largest in the world – starts operation.

The First Geophysical Congress ends in Caracas.

December 2
Petróleos de Venezuela agrees to a downstream venture with Veba Oel of West Germany.

December 9
The National Congress receives a draft bill from the Federal Executive to allow Petróleos de Venezuela to participate in Latinoil, with Pemex (Mexico) and Petrobras (Brazil).

December 13
The National Executive orders Petróleos de Venezuela to acquire US$174 million of public debt bonds.

1983

January 1
The operation areas to subsidiaries of Petróleos de Venezuela are redistributed.

Hefty reduction of investments in the Orinoco Belt field. The Jobo and Cerro Negro pilot programmes will continue.

February 18
The Government ends the fixed exchange rate system and free convertibility of the bolívar, due to flight of capital and loss of international reserves.

February 20
The Central Bank suspends foreign exchange operations.

February 22
A Presidential Decree authorises the Finance Ministry to restrict the free convertibility of the bolívar.

February 23
Lagoven completes Lorán-1, as a new field wildcat, a giant natural gasfield is discovered. (*Colón sub-basin, 90 km southeast of Galeota Pt, Trinidad, 170 km due east of the Pedernales field.*)

February 27
A Presidential Decree creates the Office for Differential Exchanges – RECADI.

Lagoven

Fig. 31. Flexicoker in Amuay refinery.

March
Lagoven starts operating its first drilling rig made in Venezuela.

March 14
A meeting of OPEC ends in London. For the first time, it is

recommended to decrease reference prices by 15%, as production quotas are also reduced.

March 23
Petróleos de Venezuela asks the Federal Executive for an end to the interference by the Central Bank in internal financial matters.

April 21
Petróleos de Venezuela and Veba Oel of West Germany sign agreements for 35 000 tonnes of crude oil/day, equal participation in the Scholven and Horst refineries near Düsseldorf, and joint research with Lurgi on heavy and extra heavy crude oil processing.

April 23
Lagoven completes Cocuina-1 as a new natural gasfield is discovered (*Colón sub-basin, 10 km northeast of the Lorán field*).

May
There is a public debate on the acquisition of German refineries by Petróleos de Venezuela.

May 6
The Eighteenth Assembly of ARPEL ends in Asunción, Paraguay.

May 31
An Extraordinary Stockholders Meeting of Petróleos de Venezuela sharply reduces the current year's budget, due to the general financial situation.

June
For the first time in years, the consumption of products decreases.

July 17
Presidents Luis Herrera-Campíns of Venezuela and Miguel de la Madrid of Mexico sign an agreement in Cancún pledging the continuation of the San José oil supply pact to Central American and Caribbean countries, as of 3 August.

July
Petróleos de Venezuela decides to liquidate Foninves (a fund for

hydrocarbons research and personnel) and INAPET (the Institute of Training and Development), and establish CEPET in their place to be in charge of human resources.

August 4
The first shipment to the USA of unleaded motor gasoline leaves the El Palito refinery.

August
Ministers of Oil from Mexico, Trinidad and Tobago, Ecuador and Venezuela end a 2-day meeting in Puerto La Cruz.

August
The Presidents of the operating subsidiaries of Petróleos de Venezuela denounce 13 executives for selling advance information on products and crude oil prices.

August 10
The HDM plant in Cardón is put on stream, to eliminate sulphur and vanadium in heavy crude oils.

August 23
A new collective contract is signed by the oil workers' unions and Petróleos de Venezuela.

August 31
President Herrera appoints Minister Calderón Berti to the post of President of Petróleos de Venezuela. The new Minister of Energy and Mines is José Ignacio Moreno-León.

September 25
Lagoven ends its exploration drilling on the Atlantic platform, on the abandonment of Tajalí No. 1 (*Colón sub-basin, 20 km west of the Lorán field, just south of the South Galeota field in Trinidad and Tobago*).

October 2
For the first time, the Government increases the price of heavy and extra heavy crude oils, independently of lighter oils.

The Yagua distribution plant in Carabobo State, 50 km south of the El Palito refinery, is inaugurated.

November 15
Plastilago goes on stream, a joint company for the production of high-density polyethylene at the El Tablazo petrochemical complex.

December 4
Jaime Lusinchi and the AD opposition party sweep the national elections.

December 9
The Sixty-Ninth OPEC Conference in Geneva resolves to maintain, unchanged, Member Country quotas and reference prices.

December 13
The Ministry of Energy and Mines authorises Maraven to start the development of the San Diego main producing area in the Orinoco Oil Belt field.

December 28
Presidential Decree 2384 creates a Commission of Oil and Petrochemical Goods and Services.

December 31
For the first time in 30 years of operation, the petrochemical company Pequiven shows a net profit.

1984

January 14
After a record 416 days of uninterrupted operation, the flexicoker unit at the Amuay refinery is shut down for cleaning and turning-up.

February 6
A new administration takes over. Jaime Lusinchi becomes President and Arturo Hernández-Grisanti Minister of Energy and Mines.

February 8
Brígido Natera is appointed President of Petróleos de Venezuela.

March 1
Petróleos de Venezuela announces that the final count on crude oil in the Orinoco Belt field is 190 million m^3, of which 32 million m^3 are recoverable by primary mechanisms due to high mobility.

February 24
The Federal Executive orders the Central Bank to purchase all foreign currency of Petróleos and its subsidiaries at Bs5.9926 per US$1.

February 25
The Second Natural Gas Workshop ends in Tamare.

March 16
There is a 4-fold increase in consumer tax on motor gasolines.

April
In relation to the renegotiation of foreign debt, the Finance and Energy Ministries form a Technical Committee to collect income tax from former concessionaires that has been outstanding for 15 years.

Corpoven discovers the Guafita oilfield (*Apure sub-basin, 45 km southwest of Guasdualito, 3 km north of the Arauca river on the Venezuelan–Colombian border, and 25 km north of the Colombian Caño Limón field*).

Pequiven

Fig. 32. Olefins plant in El Tablazo petrochemical complex.

May 15
Only two types of lubricants shall be sold in the inland market, at triple the price.

June
The first drilling barge built in Venezuela is put into service at Lake Maracaibo by Lagoven.

June 12
The Latin American oil producers end their second meeting in Port of Spain, Trinidad.

September
An OPEC delegation visits Caracas as part of a worldwide effort to obtain recognition of the need to maintain oil prices.

Corpoven discovers the La Victoria oilfield (*Apure sub-basin, 45 km west of the Guafita field*).

September 7
Coastal Corporation of Houston and Mobile Bay Refining sue Petróleos de Venezuela for US$450 million, because of its refusal to accept a reduction in the supply of crude oils in line with OPEC resolutions on production quotas.

October 11
ARPEL ends a meeting in Caracas to discuss the problems of heavy oil production and processing.

October 26
An OLADE meeting in Caracas on the perspectives and experiences in energy cooperation in Latin America ends.

October 31
The Second Latin American Energy Seminar, sponsored by the Latin America Engineers Association, ends in Caracas.

November
The President of Petrocanada ends a visit to Venezuela.

Carbozulia receives 14 proposals on an international tender for exploitation of the Guasare coal mines (*Maracaibo basin, 70 km northwest of Maracaibo*).

Petróleos de Venezuela approves the construction of the 802-km Nurgas natural gasline from the Maturín basin fields to the Paraguaná refineries.

November 13
A seminar on negotiation and the writing of contracts of technology transfer, jointly sponsored by the Foreign Investments Agency SIEX and the UN Agency for Development and Industry, ends in Caracas.

November 22
The President of the Dutch Antilles is unsuccessful in persuading Petróleos de Venezuela to take over the operation of the Aruba refinery, scheduled to be closed by Exxon in 4 months.

The Second Venezuelan Geophysical Congress ends in Caracas.

November 23
A 5-day seminar at the Latin American Centre for Studies and Research on Democracy, Socialism and Popular Power on 'Libya, the country that sowed petroleum', ends in Caracas.

November 24
The First Refining Seminar of the National Industry ends in Valencia.

November 29
Corpoven completes new field wildcat SLB-6-IX at a total depth of 5822 m, the deepest in Venezuela (*Maracaibo basin, south Lake area, block B of service contracts*).

November 30
The seventh 4-year National Plan is submitted to Congress for approval.

1985

January
Minister Hernández-Grisanti denounces the international campaign against OPEC pricing policies.

Two MPs accuse Union Carbide of financial losses to Petróleos de Venezuela, if their Puerto Rico refinery is closed down.

January 30
The Seventy-Third OPEC (Extraordinary) Conference ends in Geneva, with three Member Countries unhappy with the drop in price for light crude oils.

February
The Third Gas Workshop, co-sponsored by American Gas Processors, ends.

February 27
A seminar on the future demand for energy, co-sponsored by the European Economic Community and the Institute of Petroleum of Simón Bolívar University, ends at the Caracas Institute for Advanced Studies.

March
A visit by the Chinese Deputy Prime Minister to Venezuela coincides with the return of a high-level Venezuelan technical team from China which discussed collaboration on research on very heavy crude oils.

March 18
The Presidents of four operating companies of Petróleos de Venezuela take part in a television programme beamed to the LSU campus in Baton Rouge, Louisiana, from Caracas.

March 29
With the Ambassadors of seven OPEC Member Countries attending, a seminar on intra-OPEC trade, sponsored by the Centropep Institute, and backed by the Ministry of Energy and Mines and Caracas Central University, ends in Caracas.

April
The President of the development corporation Corpozulia proposes that the Guasare coal mines should be entrusted to Petróleos de Venezuela.

April 11
Colombia renews its contract to buy 3000 ton/day of Venezuelan crude oil during the remainder of the year.

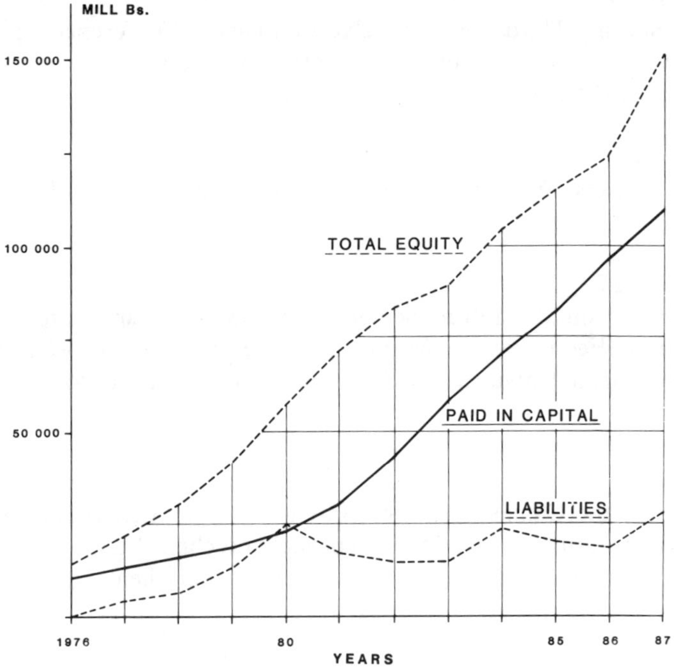

Fig. 33. Petróleos de Venezuela: total equity; paid in capital; liabilities.

April 27
An international seminar on the future of OPEC ends in Caracas.

May
The Federal Government entrusts to Petróleos de Venezuela the exploitation of large coal deposits in the Guasare river valley (*Maracaibo basin, 70 km northwest of Maracaibo*).

The General Attorney of the Republic rules that exploitation of refinery asphalt is an activity reserved to the State.

May 8
Minister Hernández-Grisanti accuses the UK and the USSR of selling large quantities of crude oil in the spot market with the purpose of disrupting the work of OPEC.

May 15
In Caracas, Venezuela and Algeria sign a protocol for cooperation and mutual aid.

May 16
The President of Algeria visits Venezuela.

May 29
Presidential Decree 645 creates the National Commission of Oceanography.

May 31
The first regional seminar for the protection of Lake Maracaibo ends in Maracaibo.

June
The second alternate injection plant for heavy crude oil along the eastern border of the Bolívar Coastal field goes into operation (*HH-8, Bachaquero area*).

June 3
A meeting of the OPEC Ministers concludes in Taif, Saudi Arabia.

June 10
Venezuela and the USSR agree to double the volume of crude oil sent by Venezuela to Cuba in exchange for Black Sea crude delivered to the Ruhr Oel refinery in West Germany, jointly owned by Petróleos de Venezuela and Veba Oel.

June 17
Shell suspends buying Venezuelan crude oils for its Curaçao refinery, which is scheduled to shut down in 1 month.

June 18
Shell asks for a 'protection price' on Venezuelan supplies of crude oil for its Curaçao refinery, in view of its imminent shut down.

June 21
The National Council for the Development of the Nuclear Industry is disbanded.

July 7
An OPEC Consultative Ministerial meeting ends in Geneva.

July 23
A team from Petróleos de Venezuela takes charge of Carbozulia.

July 25
The Seventy-Fourth OPEC Conference ends in Geneva with a drop in the price of crude oil marker.

July 27
Petróleos de Venezuela rejects Shell's plan for the operation of the Curaçao refinery.

August 3
On schedule, Venezuela and Mexico renew for the sixth year their San José agreement on oil supplies to qualifying Central American and Caribbean countries.

September 24
The statutes of Petróleos de Venezuela are revised to allow for the exploitation of 'any fossil mineral resources', i.e. Guasare coal.

September 25
Petróleos de Venezuela rents the Curaçao refinery for 5 years. A new subsidiary, Isla, is organised.

October
Lagoven start two production pilot projects (*Cerro Negro main producing area, Orinoco Belt field*).

October 1
The Curaçao refinery is shut down by Shell.

October 3
The Sixth Venezuelan Geological Congress ends in Caracas.

October 4
Venezuela asks the OPEC Conference in Vienna to start an in-depth evaluation of current policies and possibilities for the future.

Lagoven

Fig. 34. Jobo steam injection pilot plant.

October 9
A Petróleos de Venezuela Workshop entitled 'Productivity, Key to Success', ends in Maracaibo.

November 12
The Premier of China ends his visit to Venezuela – a protocol on petroleum cooperation (joint research, technical assistance, mutual training and seminars) is discussed.

November 16
A 2-day meeting of experts of GIPLACEP, the Informal Group of Latin and Caribbean Oil Exporters, ends.

November 18
The President of Ecuador ends his visit to Venezuela – OPEC strategies and research exchanges are discussed.

November 22
Meneven's cryogenic complex goes on stream in Jose to process 23 million m^3 of natural gas per day and export 9000 m^3 liquids per day.

The First Venezuela–Canada technical meeting on energy ends.

December 9
The Seventy-Sixth OPEC Conference ends in Geneva, presided over by Minister Hernández-Grisanti.

December 31
In 10 years, Petróleos de Venezuela's sales total Bs663 billion, of which exactly two-thirds is paid to the Government in taxes. The total foreign exchange earned is US$125 billion.

The Ministries of Finance and Energy and Mines set the export value of crude oil and products at 125% of the realisation price, for income tax purposes.

1986

January 1
The exchange rate of the oil dollar (US) is set at Bs7·50.

Tax on motor gasolines sold in the domestic market increases by 25% to 70%.

January 22
The Federal Executive authorises Petróleos de Venezuela to acquire 50% of Nynas Petroleum (Sweden) and to increase its shares in the Veba Oel refineries in West Germany.

February
Lagoven discovers a giant field in a new light crude oil trend, on completion of El Furrial No. 1 (*Maturín basin, 17 km west of Maturín*).

February 5
Petróleos de Venezuela signs a letter of intention to buy 50% of Citgo Petroleum, a subsidiary of Southland Corporation of Dallas.

February 16
Maraven completes Tomoporo-1, a new field discovery well (*Maracaibo basin*).

February 19
Petróleos de Venezuela files revised income tax estimates, due to a drop in international oil prices.

February 24
Interamerican Bank gives a US$100 million loan to Venezuela, the first after the Third World debt crisis.

February 26
The Minister of Finance signs refinancing agreements in New York with 460 foreign banks for US$21 billion foreign debt.

March
The cost of producing 1 m^3 of Venezuelan oil is US$22 (US$3.50 = 1 barrel).

March 11
The price of crude oil and its products is set at 120% realisation for income tax purposes.

March 24
The first session of the Seventy-Seventh OPEC (Extraordinary) Conference ends in Geneva.

Fig. 35. El Furrial No. 1 is the first successful completion in a deep trend along the northern part of the Maturín basin. As of year-end 1987, four other supergiant oilfields have been discovered.

March 31
Petróleos de Venezuela's capital is increased to Bs95 billion during the annual stockholder's meeting.

April
Corpoven starts construction of the 802-km Nurgas gasline from the Anaco fields to Paraguaná and Zulia.

April 2
By Presidential Decree, investments in the exploitation, industrialisation and marketing of coal and petrochemical products are to be exempt of tax for 5 years.

April
President Lusinchi announces in Brasilia that Venezuela will buy 30 Tucano trainer planes with crude oil.

April 11
A Venezuela–USA technical meeting on petroleum information ends in Caracas.

April 12
Presidents Lusinchi of Venezuela and Raúl Alfonsín of Argentina end 3 days of meetings in Buenos Aires, referring in passing to the need that all oil producers, whether a member of OPEC or not, ensure stable prices in the international market.

April 14
Petróleos de Venezuela signs letters of intention with the Union Pacific Corporation to buy 50% of the Tulsa based Champlin Petroleum.

April 21
The Seventy-Seventh OPEC (Extraordinary) Conference ends in Geneva. Ten Member Countries agree on realistic production quotas in order to stabilise prices.

April 28
Petróleos de Venezuela buys all shares of Carbozulia; those of the Venezuelan Fund for Bs77 million; those of the Zulia Development Corporation, Corpozulia, for Bs100 million.

May 5
The Prime Minister of Trinidad and Tobago ends a 3-day visit to Venezuela. Talks are centred on extra-OPEC cooperation and an increase in binational commercial oil relations.

May 29
Grupo Químico, a company of the Venezuelan Mendoza group, puts on stream a chemical lubricants plant.

June
Blow-out of new pool wildcat SLB-5-4X (*Maracaibo basin, in block B of former Shell Surca service contract area*).

The largest oil spill is cleaned up, with more than 150 km of coastline affected (*Bachaquero area, Bolívar Coastal field*).

June 1
Petróleos de Venezuela merges Meneven with Corpoven, and creates Interven to handle internationalisation.

June 3
The Ministry of Foreign Affairs of Colombia acknowledges Venezuelan sovereignty over the Los Monjes archipelago (*bordering the Gulf of Venezuela basin*).

June 9
By Presidential Decree, Petróleos de Venezuela is appointed general co-ordinator and executor of the National Contingency Massive Oil Spills Plan.

June 12
The Fifth Latin American Oil Exposition closes in Maracaibo.

President Lusinchi orders that all legal proceedings against former concessionaires, brought by the Comptroller General for tax claims, should be transacted.

June 21
The Prime Minister of Norway, after a 3-day visit, announces eventual cooperation with OPEC.

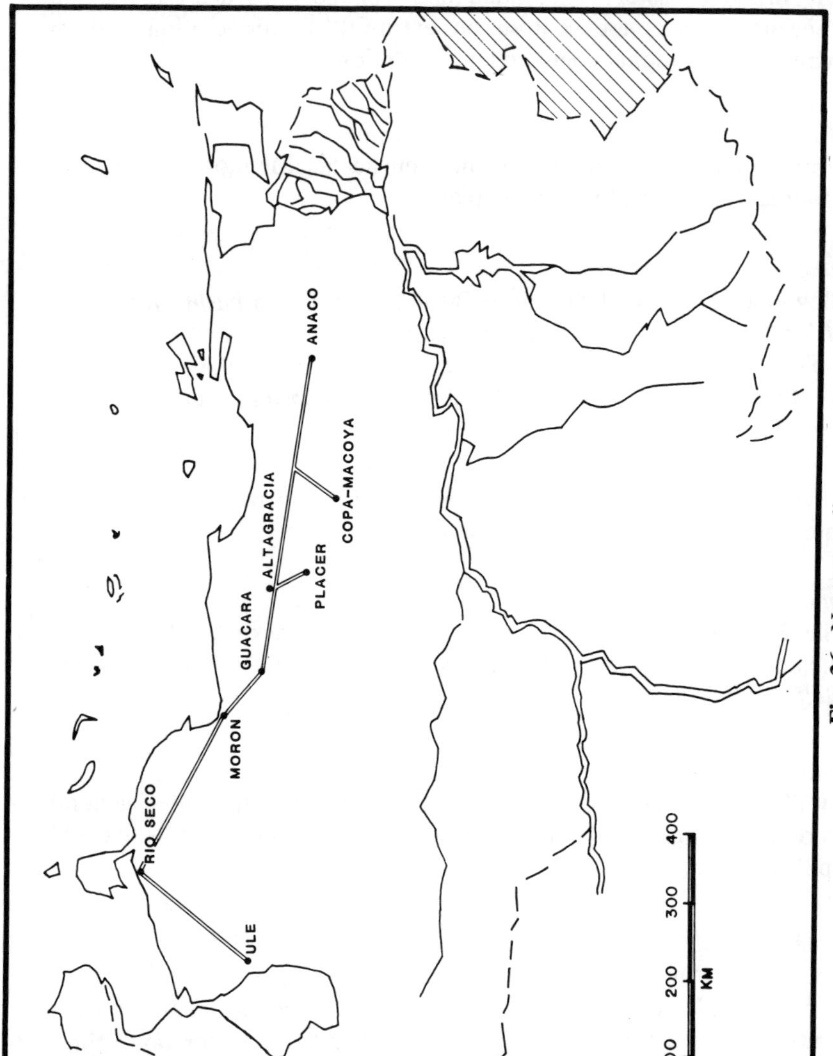

Fig. 36. Nurgas gasline.

June 23
The Ministry of Energy and Mines announces regulations for the transport, storage and selling of petroleum products in the domestic market.

June 30
Petróleos de Venezuela agrees with the Axel Johnson group of Sweden to buy 50% of Nynas Petroleum.

The first session of the Seventy-Eighth OPEC Conference ends in Brioni, Jugoslavia. The basis for cooperation between OPEC and non-OPEC oil-producing and exporting countries was examined.

July 1
President Lusinchi sets the value of the Venezuelan bolívar at Bs7·50 equals US$1.

August
Lagoven and New Brunswick Power (Canada), commence a pilot programme on the use of Orinoco Belt natural bitumen-in-water emulsion for direct burning for electricity generation, following research in Intevep (*Orimulsión*).

August 3
The seventh renewal of the San José agreement between Venezuela and Mexico takes place.

August 5
New regulations are set by the Ministry of Energy and Mines on the use of natural gas in the domestic market.

The Seventy-Eighth OPEC Conference ends in Geneva. A ceiling on production is continued, Iraq excepted.

September 5
A decrease in oil income of US$5 billion, due to a drop in prices, is estimated by Minister Hernández-Grisanti at the annual stockholder's meeting of Petróleos de Venezuela.

September 10
The Federal Executive authorises Petróleos de Venezuela to acquire 50% of Citgo Petroleum.

September 12
Coastal and Mobil Bay withdraw a law suit against Petróleos de Venezuela.

September 13
On the wake of the first visit by a Venezuelan President to Trinidad and Tobago, it is agreed to increase cooperation in energy matters. The Trinidadian Government will actively participate in the massive oil spills contingency plan.

September 14
Corpoven completes Musipan-1 as a new field discovery, and at a depth of 5420 m, it is the deepest in eastern Venezuela (*Maturín basin, 15 km west of the El Furrial field*).

September 15
Notwithstanding a public outcry, the National Congress votes in favour of claims for Bs4·5 billion against former concessionaires being dropped, in exchange for the non-payment of Bs1·7 billion due to the concessionaire companies under the Guarantee Fund, created when the oil industry and commerce were reserved to the State in 1975.

September 19
The Ministry of Energy and Mines fixes the geographical distribution of responsibilities among the operating subsidiaries of Petróleos de Venezuela.

September 26
The Third Geophysical Congress ends in Caracas.

September 30
Pablo Reimpell is elected President of Citgo Petroleum.

October 22
The Seventy-Ninth OPEC (Extraordinary) Conference ends in Geneva. Production interim quotas are set.

December 20
The 9-day Eightieth OPEC Conference ends in Geneva. Prices are set based on a basket of seven crude oils.

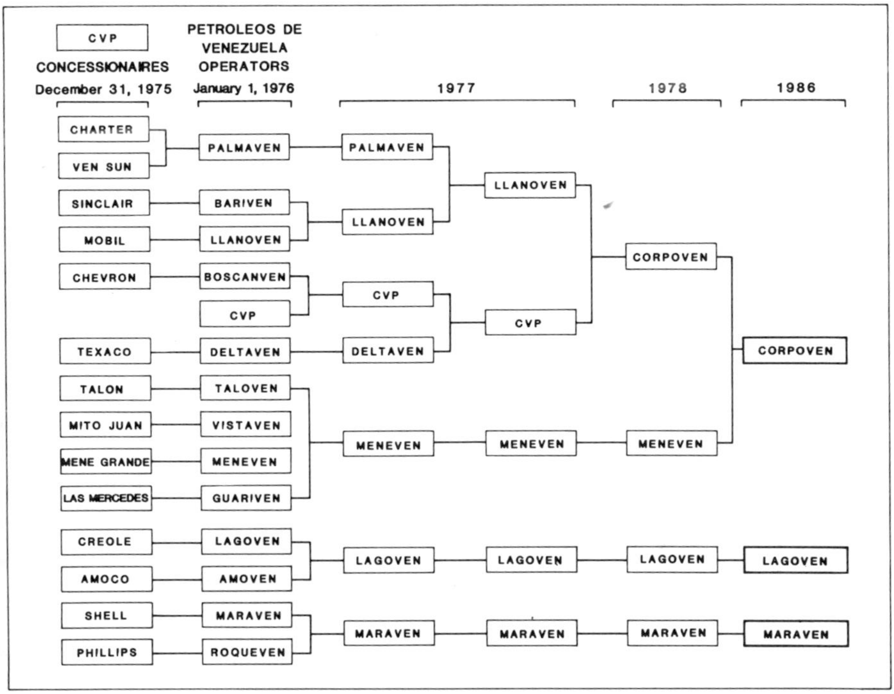

Fig. 37. Concessionaires; operators.

1987

January 8

Pequiven signs an agreement for a joint venture to produce polypropylene in El Tablazo.

February

Maraven starts a steam injection project at a depth of 3000 m in the Boscán field.

February 1

OPEC Member Countries notify an increase in prices of crude oil on the basis of a basket of seven types.

February 26
Petróleos de Venezuela agrees with the National Academy of History a programme for the recuperation and conservation of historical documents.

February 27
An International Symposium on enhanced recovery SIREMCRU II ends in Maracaibo.

March 8
Venezuela agrees to send 1 million tonnes of crude oil to Ecuador, to compensate losses due to the destruction of a pipeline by an earthquake.

March 10
The new president of Veba Oel visits Caracas in connection with a meeting of the stockholders of Ruhr Oel.

March 11
The National Government agrees to the acquisition by Petróleos de Venezuela of a 50% interest in Champlin Petroleum.

Export prices of Venezuelan crude oils are increased to 17%, backdated to 1 January.

March 17
OPEC and the EEC end their first technical seminar on the medium- and long-term outlook for energy in Luxembourg.

Petróleos de Venezuela buys 50% of Champlin Petroleum, a refining company based in Corpus Christie, Texas.

March 23
Pequiven and Ecofuel, a subsidiary of ENI, sign an agreement to form Super Octanos, a joint venture to build and operate an MTBE plant in Jose.

March 25
Carbozulia agrees to constitute a joint company with ACC Venezuela (Arco) and AGIP Carbone to exploit and market the Paso Diablo coal mine (*Maracaibo basin, 70 km northwest of Maracaibo*).

March 27
The Fifth Ministerial Meeting of GIPLACEP ends in Caracas.

April 1
Petróleos de Venezuela (UK), with its headquarters in London, becomes Petróleos de Venezuela (Europe).

April 12
The Fifth Workshop of GPA (Venezuela) on natural gas ends in Maracaibo.

May 8
Venezuela and Mexico sign a Technical Cooperation Agreement in Mexico City.

May 9
The Twenty-Second General Assembly of ARPEL ends in Caracas.

June 16
The First Symposium on the Production Geology of Petróleos de Venezuela and its operating subsidiaries ends in Puerto La Cruz.

June 17
The Eighth Ministerial Meeting of the OPEC Fund ends in Caracas.

June 27
The Eighty-First OPEC Conference in Vienna limits production with a view to strengthening prices.

June 29
The Federal Executive sets the oil dollar (US) exchange rate at Bs14·50.

July 7
Minister Hernández Grisanti signs an agreement to implement joint technical research and development projects on petrochemicals with the President of the Petrochemical Corporation of China in Beijing.

July 30
President Lusinchi of Venezuela and the President of Mexico renew the San José agreement in Mexico City.

August 12
Petróleos de Venezuela and Elf-Aquitaine sign a cooperation agreement on research and technological services.

September
Maraven starts geological exploration of the northwestern Andean region (*Maracaibo basin*).

September 15
Well MFB-293 (*Machete main producing area, Orinoco Belt field*) following steam injection, tests more than 1000 m^3/day.

September 22
The first shipment of Guasare coal by Carbozulia takes place.

October 15
The Ministry of Energy and Mines approves a 54-km pipeline from the San Diego main producing area of the Orinoco Belt field to the Budare field, to test the movement of extra heavy oil by core flow, a system developed by Intevep.

October 16
Petróleos de Venezuela and Brazil sign a cooperation agreement in the oil and petrochemical industries.

November
New Brunswick Electric Power successfully completes a test for the burning of new fuel Orinoco natural bitumen emulsion in water, directly for the generation of electricity.

November 1
In Willemstadt, President Lusinchi signs a document extending the lease of the Curaçao refinery to 9 years.

November 9
The Price Commission of OPEC visits Venezuela.

November 26
Palmaven, a subsidiary of Pequiven, in charge of the supply of fertilisers to the domestic market, is converted to a subsidiary of Petróleos de Venezuela.

Fig. 38. Production module '0-16' in Cerro Negro main producing area, Orinoco Oil Belt, provider of natural bitumen for *Orimulsion.*

December 14
The Eighty-Second OPEC Conference maintains production quotas for the first semester of the coming year.

1988

January 1
Venezuela becomes 3rd exporter of oil to the USA.

January 17
The Second Geological Workshop on Coal ends in Maracaibo.

The First Financial Seminar of Petróleos de Venezuela ends in Caracas.

January 18
Continued blow-ups of oil pipeline in Colombia from Caño Limón fields are a threat to the delicate ecological system in southwestern Venezuela, even reaching Lake Maracaibo.

February 10
Seminar on oil industry in Venezuela ends in London.

Fig. 39. Petróleos de Venezuela and affiliates.

March 8
In Caracas, UK Minister of State announces cooperation programmes in hydrocarbons with Venezuela.

Petróleos de Venezuela and Barbados National Oil Company sign a technical support agreement.

March 9
President Jaime Lusinchi's 'State of the Nation' address to the National Congress indicates that new fuel Orimulsion will be marketed worldwide shortly. Natural bitumen, being a product, is outside OPEC quotas.

March 17
The Third Symposium on Oil Exploration of Sub-Andean Basins ends in Caracas.

March 18
The capital of Petróleos de Venezuela increased to Bs128 325 million.

April
The pilot programme of direct electricity generation through Orimulsion is a success in Japan.

April 20
The USSR is ready to help the development of the Orinoco Belt field, according to an announcement by the Academician V. Volski in Caracas.

May 17
Aníbal R. Martínez proposes internationalisation of exploration.

July 1
Proposal by Colombia to build a pipeline to Lake Maracaibo from the Caño Limõn fields triggers anger and a general debate against the project.

July 9
President Desmond Hoyte of Guyana proposes a joint exploitation of any oil deposits shared with Venezuela.

July 31
After 75 years, Zumaque well No. 1 (*Mene Grande field*) is still an oil producer.

August 8
Julio César Gil becomes new Minister of Energy and Mines.

August 18
Secretary General of OPEC Subroto ends a week-long visit to Venezuela.

August 20
A new subsidiary of Petróleos de Venezuela Bitúmenes del Orinoco is created to market Orimulsion.

September 8
The Fourth International Conference on Heavy Oil and Tar Sands ends in Edmonton, Alberta, Canada.

September 15
The Fourth Geophysical Congress ends in Caracas.

September 21
The Venezuela–Canadá technical meeting on oil ends in Ottawa.

September 22
Petróleos de Venezuela and Nynas Petroleum complete an enlargement of a refinery in Sweden.

September 26
A Venezuela–USA technical meeting on oil ends in Washington.

September 30
Petróleos de Venezuela acquires a half interest in the Champlin refinery in Corpus Christie TX-US, thus becoming sole owner.

An agreement is signed in Caracas between the Ministry of Energy and Mines of Venezuela and the Ministry of Geology of China, for co-operation in hydrocarbons and mineral prospection.

October 3
Proved reserves of natural gas of Venezuela reach 5000 million m^3.

October 6
The Minister of Energy and Mines, Julio César Gil, denies that Venezuelan oil goes into the strategic reserve of the USA.

October 10
Petróleos de Venezuela is to buy refinery and service marketing outlets of Union Oil of California in the Chicago IL-US area.

Tables

TABLES

TABLE 1
Exploration (Venezuela)

Year	*Wildcats/ discoveries*	*Km seismic*	*Investments (Bs million)*
1976	24/15	5 600	400
1977	50/20	5 500	413
1978	61/42	17 600	781
1979	118/55	17 700	1 380
1980	345/90 (188 susp)	33 400	2 194
1981	309/56 (203)	12 900	2 696
1982	225/87 (89)	13 522	3 053
1983	70/23	4 391	1 687
1984	24/15	5 192	1 079
1985	31/11	16 097	1 644
1986	32/16	10 572	1 789
1987	26/9	5 750[a]	1 655

Source: Petróleos de Venezuela.
[a] 3D: 165 km^2.

TABLE 2
Crude Oil Proved Reserves by Density (Venezuela)[a]

Year	*Light*	*Medium*	*Heavy*	*Extra-heavy*	*% Heavy and % Extraheavy*
1975	487	817	1 582	38	55
1980	518	801	1 617	115	58
1981	549	821	1 091	143	57
1982	599	803	1 762	744	64
1983	638	845	1 845	828	65
1984	681	897	1 848	1 030	65
1985	733	910	1 994	1 025	65
1986	1 106	1 155	2 645	3 923	74
1987	1 107	1 340	2 680	4 116	74

Source: Ministry of Energy and Mines.
[a] Volumes in $m^3 \times 10^6$.

TABLE 3
Production (Venezuela)

Year	*Wells/ workovers*	*Production* ($m^3 \times 10^6$)	*Potential* ($m^3 \times 10^6$)	*Investments* (*Bs millions*)
1976	285/1 438	134		870
1977	539/1 386	130		1 484
1978	653/1 368	126		2 178
1979	668/1 470	137	162 (goal)	3 124
1980	656/1 659	126		4 127
1981	673/1 697	122		6 464
1982	931/1 621	110		10 322
1983	796/1 784	105		9 358
1984	568/1 739	99	150	8 846
1985	449/1 349	98	150	8 461
1986	236/1 236	104	150	8 708
1987	124/1 097	106	150	10 992

Source: Petróleos de Venezuela.

TABLE 4
Crude Oil Production and Exports (Venezuela)

Year	*Production* ($m^3 \times 10^6$)	*Exports* ($m^3 \times 10^6$)	%
1917–1919	0·12	0·03	25
1920–1924	2·8	2·3	82
1925–1929	57	54	95
1930–1934	99	95	96
1935–1939	140	131	94
1940–1944	158	150	95
1945–1949	337	325	96
1950–1954	503	475	94
1955–1959	741	691	93
1960–1964	907	856	94
1965–1969	1 021	961	94
1970–1974	978	911	93
1975–1979	662	592	89
1980–1984	563	466	83
1985	98	79	81
1986	104	88	85
1987	106	88	83
Total	6 480	5 964	92

Source: Ministry of Energy and Mines.

TABLE 5
Production and Use of Natural Gas (Venezuela)[a]

Year	*Gas production*	*Injected*	*Used*	*Lost*	*Proved reserves*
1975	38 088	20 986	12 751	4 298	1 173 614
1980	35 449	16 534	16 680	2 235	1 262 391
1981	34 681	16 119	16 642	1 920	1 365 135
1982	33 286	13 860	17 638	1 788	1 471 215
1983	31 766	12 866	17 165	1 735	1 562 332
1984	32 574	12 030	18 788	1 756	1 666 629
1985	32 996	12 428	18 869	1 617	1 754 312
1986	36 275	12 040	21 460	2 775	2 622 000
1987	36 614	11 648	21 254	3 712	2 765 000

Source: Ministry of Energy and Mines.
[a] Volumes in $m^3 \times 10^6$.

TABLE 6
Refining (Venezuela)

Year	*Processed (ton/year* $\times 10^6$*)*	*% Usage of capacity*	*Investments (Bs millions)*
1976	141	67	28
1977	138	68	163
1978	140	68	740
1979	140	64	1 820
1980	132	64	2 952
1981	123	65	2 604
1982	124	66	2 584
1983	122	68	772
1984	119	74	664
1985	130	74	640
1986	129	72	1 133
1987	130	72	n.a.

Source: Petróleos de Venezuela.

TABLE 7
Refineries (Petróleos de Venezuela)

	Volume processed (ton/year × 10^6)	
	1986	*1987*
Venezuela		
Amuay	19	20
Cardón	13	12
Puerto La Cruz/El Chaure	5	6
El Palito/El Toreño	5	5
Bajo Grande	2	2
San Roque	0·25	0·25
W. Germany		
Scholven (50%)	3	3
Horst (50%)	3	3
Neustadt (12·5%)	2	2
Karlsruhe (16%)	2	2
USA		
Lake Charles (50%)	8	8
Corpus Christie (50%)	—	4
Sweden		
Nynashamm (50%)	0·7	0·7
Gothemburg (50%)	0·3	0·3
Belgium		
Antwerp (50%)	0·4	0·4

Source: Petróleos de Venezuela.

TABLE 8
Transport by Trunk Lines (Venezuela)[a]

Year	*Crude oil*	*Natural gas*	*Products*
1975	173	8 854	12
1978	163	11 358	17
1980	158	12 991	16
1982	151	12 360	15
1984	130	14 713	21
1985	127	14 000	22
1986	140	17 760	24
1987	131		

Source: Division of Production and Conservation, Ministry of Energy and Mines.
[a] Volumes in $m^3 \times 10^6$.

TABLE 9
Domestic Market (Venezuela)

Year	*Motor gasolines (litres/day × 10^6)*	*Total (litres/day × 10^6)*	%[a]	*Investments (Bs millions)*
1976	19	39	49	
1977	21	42	50	
1978	23	45	51	92
1979	24	50	48	117
1980	25	55	45	277
1981	26	56	46	307
1982	27	58	47	309
1983	27	55	49	282
1984	26	51	51	361
1985	26	51	51	953
1986	26	53	51	1 267
1987	27	53	51	1 592

Source: Petróleos de Venezuela.
[a] Percentage of motor gasolines consumed with respect to total.

TABLE 10
Payments for Technical Support (Petróleos de Venezuela)

Year	*Payment (Bs millions)*
1975	—
1976	572
1977	681
1978	632
1979	550
1980	142
1981	120
1982	80
1983	52
1984	52
1985	—
1986	—
1987	—
Total	2 881

Source: Department of Finance and Fiscal Studies, Ministry of Energy and Mines.

TABLE 11

Contracted engineering (Petróleos de Venezuela)[a]

Year	*Venezuelan companies*	*Foreign companies*	*% Ven./For.*
1980	659	3 383	16/84
1981	1 075	2 532	30/70
1982	1 830	3 100	37/63
1983	2 280	1 230	65/35
1984	1 226	452	73/27
1985	1 632	402	80/20
1986	2 237	392	85/15
1987	2 460	475	84/16

Source: Petróleos de Venezuela.
[a] In man hours × 10^3.

TABLE 12

Staff (Petróleos de Venezuela)

Year	*Employees*	*Workers*	*Total*	*Men hours (millions)*
1974	13 079	10 573	23 652	45·2
1978	14 382	11 738	26 120	43·1
1982	22 782	16 407	39 189	67·4
1983	23 648	15 947	39 595	68·1
1984	23 655	16 450	40 105	57·2
1985	23 558	14 843	38 401	56·1
1986	24 320	14 604	38 924	66·7
1987	24 693	14 445	39 138	59·3

Source: Division of Hydrocarbons, Ministry of Energy and Mines.

TABLE 13

Sale of Petrochemicals (Pequiven)[a]

Year	*Domestic market*	*Exports*	*Joint ventures*
1981	486	481	200
1982	557	542	192
1983	629	469	219
1984	994	514	311
1985	1 220	421	307
1986	1 576	426	364
1987	1 894	481	362

Source: Petrochemicals Division, Ministry of Energy and Mines.
[a] Volumes in tonnes (000s).

TABLE 14
Joint ventures (Pequiven)

Company	*Site*	*Products*	*Capacity (ton/year)*
Nitroven	El Tablazo	Ammonia	594 000
		Urea	792 000
Plastilago	El Tablazo	High-density polyethylene	60 000
Polilago	El Tablazo	Low-density polyethylene	80 000
Estizulia	El Tablazo	Polyethylene	40 500
Química Venoco	El Tablazo	Propylene tetramer	18 500
	Guacara	Duodecylbenzene	18 000
Produven	Morón	Freons	9 900
Oxidor	Guacara	Ftalic anhydride	15 000
Ferralca	El Tablazo	Aluminium sulphate	60 000
Tripoliven	El Tablazo	Sodium phosphates	30 000
Pralca	Santa Rita	Ethylene oxide and ethyleneglycol	30 000
Superoctanos	Jóse	MTBE	500 000
Propilven	El Tablazo	Polypropylene	70 000

Source: Pequiven.

TABLE 15
Production and Investments (Carbozulia)

Year	*Production (tonnes)*	*Exports (tonnes)*	*Coal/mine development (Bs millions)*	*Capital (Bs millions)*
1986	—	—	—	450
1987	117 500	58 700	702	450

Source: Petróleos de Venezuela.

TABLE 16
Investments Abroad (Petróleos de Venezuela)[a]

Year	*Ruhr Oel*	*Citgo Pet.*	*Nynas Pet.*	*Champlin*
1985	784 (acq)[b]			
1986	802	2 175 (acq)[b]	182 (acq)[b]	
1987	827	2 756	260	232 (acq)[b]

Source: Petróleos de Venezuela.
[a] Equity value in mill Bs (millions).
[b] acq: Acquisition.

TABLE 17

Compensation Payments to Former Concessionaires and Equity Holders (31 December 1975)[a]

	Total	*Deductions*	*Payable*	*Cash*	*In bonds*
Amoco	17·82	5·08	12·74	4·04	8·70
Caracas	1·73	0·53	1·20	0·22	0·98
Chevron	14·84	2·73	12·11	0·60	11·51
Continental	1·13	0·05	1·08	+	1·07
Coro	1·46	0·28	1·18	0·16	1·02
Creole	589·13	113·55	475·58	71·47	404·11
Charter	4·38	0·38	4·00	0·86	3·14
Eastern Gas	0·08	0·01	0·07	0·00	0·07
International	53·84	16·96	36·88	1·42	35·46
Mene Grande	105·38	17·87	87·51	2·90	84·61
Mito Juan	4·13	0·65	3·48	0·13	3·35
Mobil	31·10	8·07	23·03	7·63	15·40
Philips	25·60	5·70	19·90	3·51	16·39
Las Mercedes	7·44	5·16	2·28	0·16	2·12
Shell	328·81	79·01	249·80	10·71	239·09
Sinclair	9·07	2·62	6·45	1·19	5·26
Talon	5·10	3·12	1·98	0·97	1·01
Texaco Maracaibo	37·48	6·53	30·95	0·57	30·38
Texas	14·98	5·02	9·96	1·21	8·75
Venezuelan Atlantic	20·55	2·73	17·82	1·94	15·88
Vengref	8·39	2·76	5·63	1·27	4·36
Venezuelan Sun	31·67	4·30	27·37	3·27	24·10
	1 314·11	283·11	1 031·00	114·23	916·76
Ashland	2·40	0·74	1·66	0·22	0·44
Petrofina	0·04	0·01	0·03	+	0·03
Guanipa	0·41	0·06	0·35	–	0·35
Murphy	1·76	0·35	1·41	0·16	1·25
Bajomar (Monsanto)	0·60	0·03	0·57	+	0·57
Petrobelge	0·04	0·01	0·03	+	0·03
Pure	4·33	0·36	3·97	0·86	3·11
Sunray	5·68	1·33	4·35	0·80	3·55
Triangle	1·59	0·19	1·40	0·23	1·17
Texaco Seaboard	5·54	0·96	4·58	0·34	4·24
Tenneco	0·62	0·05	0·57	–	0·57
Ucar	3·19	0·63	2·56	0·28	2·28
Unión Petrolera	0·59	0·05	0·54	–	0·54
Venezoil	0·01	—	0·01	+	0·01
Venezuelan Canadian	0·41	0·08	0·33	0·08	0·25
Venezuelan Pacific	0·96	0·19	0·77	0·18	0·59
	28·17	5·04	23·13	3·15	19·98
Total	1 342·28	288·15	1 054·13	117·38	936·74

Source: Official Gazette Extr. 1784, 18 December 1975.

[a] In US$ millions.

TABLE 18

Income Tax and Comptroller's Claims Transactions with Former Concessionaires (15 September 1986)[a]

Company	*Amount*	*Transacted for (in Guarantee fund)*	*Amount (% of total claim)*	*Transaction (% of amount)*
Creole Petroleum	1 722	620	38	36
Shell de Venezuela	1 364	549	30	40
Mene Grande	276	89	6	32
Mobil	237	49	5	21
Chevron	149	22	3	15
Texaco Maracaibo	94	23	2	24
Amoco	84	10		12
Venezuelan Sun	79	22		28
Texas Petroleum	61	47		77
Venezuelan Atlantic	56	25		45
International Petroleum	46	0		—
Philips	37	27		73
A. Johnson & Co.	33	0		—
Venezuela Gulf	29	13		45
Sinclair	28	21		75
Ucar Interam	26	2		8
Internat. Pet. (Venezuela)	23	47		(204)
Shell International	18	0		—
Texaco Seaboard	18	7		39
Scallop Corporation	17	0		—
Sunray	15	4		27
Talon Petroleum	11	0·5		5
Ultramar	10	4		40
34 others				
Total	4 500	1 614[b]		36

Source: Official Gazette 33 564, 26 September 1986.
[a] In Bs millions.
[b] 1971, US$375 million; 1986, US$111 million.

Bibliography

BIBLIOGRAPHY

GENERAL

A. ROJAS *Primer libro de geografía de Venezuela según Codazzi.* Caracas: Rojas Hnos., 1870.
M. TEJERA *Venezuela Pintoresca e ilustrada.* París: Lib. Española (E. Denné Schmitz), 1875.
W. SIEVERS *Venezuela.* Hamburg: Friederichsen, 1888.
C. RICHARDSON *The modern asphalt pavement.* London, 1908.
L.V. DALTON *Venezuela.* London: Fisher and Unwin, 1912.
C.C. McDERMOND *Who's who in Venezuela.* Maracaibo: McDermond, 1932.
MINISTERIO DÈ FOMENTO *Boletín* (Sup. N°. 1, 30 Jun.; No°. 2, 31 Oct.). Caracas, 1936.
P.H. GIDDENS *The birth of the oil industry.* New York: Macmillan, 1938.
M.W. BALL *This fascinating oil business.* Indianapolis: Bobbs-Merril, 1940.
B.T. BROOKS. *Peace, plenty and petroleum.* Lancaster (Pa): Jacques Catlett Press, 1944.
L.M. FANNING *Our oil resources.* New York: McGraw-Hill, 1945.
H.D. HEDBERG and F. HEDBERG *Bibliografía e índice de la geología de Venezuela.* Caracas: Lit. del Comercio, 1945.
R.A. LIDDLE *The geology of Venezuela and Trinidad.* Ithaca (NY): Paleont. Research Inst., 1946.
M.R. EGAÑA *Tres décadas de producción petrolera.* Caracas: Tip Americana, 1947.
H. PITTIER *Trabajos escogidos.* Caracas: Min. Agric. and Cría, 1948.
J. BESSON *Historia del Estado Zulia,* Maracaibo: Hnos. Belloso Rosell; Vol. I, 1945; Vol. II, 1948; Vol. III, 1949; Vol. IV, 1951; Vol. V, 1957.
T. POLANCO MARTINEZ *Esbozo sobre historia económica venezolana (1948–1810).* Caracas: Ancora, 1950.
ILO (International Labour Office) *Freedom of association and conditions of work in Venezuela.* Geneva, 1951.
A.D. AGUERREVERE *Elementos de derecho minero.* Caracas: Ragón, 1954.

R.W. HIDY and M.E. HIDY *Pioneering in big business*; *The resurgent years*. New York: Harper and Bros., 1955, 1956.

E. LIEUWEN *Petroleum in Venezuela*. Berkeley: University of California Press, 1955.

R. BETANCOURT *Venezuela: política y Petróleo*. México: Fondo Cult. Econ., 1956.

MMH (Ministry of Mines and Hydrocarbons) *Stratigraphical Lexicon of Venezuela*. Caracas: Sucre, 1956.

VOSA (Venezuelan Oil Scouting Association) *Names of 300 oil companies registered in Venezuela, 1920–1955*. Caracas, 1957.

R. GONZALEZ MIRANDA *Estudios acerca del régimen legal del petróleo en Venezuela*. Caracas: Fac. Derecho, Central University, 1958.

HNO. N. MARIA *Los Orígenes de Maracaibo*. Maracaibo: University of Zulia, 1959.

R. ARNOLD, G. MACREADY and TH. BARRINGTON *The first big oil hunt*. New York: Vantage Press, 1960.

W.D. MARSLAND and A. MARSLAND *Venezuela through its history*. New York: Crowell, 1961.

MMH (Ministry of Mines and Hydrocarbons). *Venezuela and OPEC*. Caracas: Imp. Nacional, 1962.

M.M. RAMOS *La industria petrolera en las estampillas de correos*. Rev. Shell, Dic., 1962.

J.P. PEREZ ALFONZO *Petróleo, jugo de la tierra*. Caracas: Arte, 1961.

J. PRIETO SOTO *El Chorro: gracia o maldición*. Maracaibo: University of Zulia, 1962.

S. DE LA PLAZA *Desarrollo económico*. Caracas: Central University, 1962.

J.A. CLARK *The chronological history of the petroleum and natural gas industry*. Houston: Clark Book Co., 1963.

A.R. MARTINEZ *Nuestro Petróleo*. Madrid: Gráficas Minerva, 1963.

VOSA (Venezuelan Oil Scouting Association). *Catalogue of Venezuelan Oilfields and wells, 1878–1961*. (*Supplements* for 1962, 1963, 1964, 1965, 1966). Caracas, 1963.

G. MORON *History of Venezuela*. London: Allen and Unwin, 1964.

C. ROMERO and L.G. ARCAY *Indice de leyes vigentes*. Caracas: Mene Grande Oil Co., 1966.

C. BALESTRINI *La industria petrolera en Venezuela*. Caracas: Ministry of Mines and Hydrocarbons, 1966.

BCV (Central Bank) *La economía venezolana en los últimos venticinco años*. Caracas: BCV, 1966.

CREOLE (Creole Petroleum Corp.) *Historia de la industria petrolera en Venezuela*. (F.B. Baptista). Caracas, 1966.

CREOLE *Temas petroleros*. Caracas, 1966.

A.R. MARTINEZ *Our gift, our oil*. Dordrecht (Holland): Reidel, 1966.

R. SADER PEREZ *Petróleo nacional y opinión pública*. Caracas: Ofidí, 1966.

M. EL-SAYED *L'organization des pays exportateurs de petróle*. París: Ed. Lib. Gen. Droit and Jurisp., 1967.

A. SIVOLI *Venezuela y sus riquezas mineras.* Caracas: Cuatricentenario, 1967.
F. GUERRERO MATHEUS *En la ciudad y el tiempo.* Maracaibo: Bco. Fom. Reg. Zulia, 1967.
D. BENDAHAN *La legislación venezolana sobre hidrocarburos.* Caracas: Emp. El Cojo, 1969.
G.J. SALAS *Petróleo.* Caracas: Monte Avila, 1969.
L. GONZALEZ BERTI *Compendio de derecho minero venezolano.* Pub. Fac. Derecho. ULA, Mérida. Vol. I, 1969; Vol. II, 1970.
F. ALVAREZ CHACIN *Secretos petroleros contra Venezuela.* Caracas: La Bodoniana, 1970.
A. ARELLANO MORENO *Mensajes presidenciales.* (Prólogo). Vol. I. Caracas: Pres. Rep. 1970.
D. BENDAHAN *Manual de legislación laboral venezolana.* Caracas: Sucre, 1971.
L. GONZALEZ BERTI *Contratos de servicio y nuevos aspectos impositivos.* Mérida: ULA, 1972.
P.E. MEJIA ALARCON *La industria del Petróleo en Venezuela.* Caracas: Central University, 1972.
E.W. OWEN *Trek of the oil finders.* American Assoc. Pet. Geol., Tulsa USA, 1975.
A.R. MARTINEZ *Gumersindo Torres.* Caracas: Pres. Rep., 1975.
I. RODRIGUEZ GALLAD and F. YANEZ *Cronología ideológica de la nacionalización petrolera en Venezuela.* Caracas: UCV, 1977.
CONICIT *Fuentes Energéticas: Una perspectiva venezolana.* Caracas: A.R. Martinez, 1977.
O. GERSTL *Memorias e historias.* Caracas: Fund, Boulton, 1977.
B.S. McBETH and W.M. SULLIVAN *Historia Documental de la Industria Petrolera.* Caracas: Funres, 1978.
A.R. MARTINEZ *La diferencia con Colombia.* Caracas: Edreca, 1982.
G. CORONEL *The Nationalization of the Venezuelan Oil Industry.* Lexington: Lexington Books, 1983.
E.B. BROSSARD *Petroleum: Politics and power.* Tulsa, USA: Penwell, 1983.
J.C. ARREAZA *Petróleos de Venezuela: Une organización eficiente y productiva.* Caracas, 1983.
A.R. MARTINEZ *Biografía de Juan Pablo Pérez Alfonzo.* Col. Venezolanos del S. XX. Fund. "E. Mendoza", Vol. 3, pp. 66–84. Caracas, 1984.
J. EWELL *Venezuela: Century of change.* Stanford University Press, 1984.
ULTRAMAR PLC. *A golden adventure* (The first 50 years of Ultramar). London; Hurtwood, 1985.
PETRÓLEOS DE VENEZUELA *1976–1985: Diez años de la industria petrolera nacional.* Vol. I: J. Giacopini-Zárraga 'Antecedentes históricos', G. Rodríguez-Eraso 'Aspectos operacionales y administrativos'; Vol. II: J.C. Arreaza 'Aspectos históricos y jurídicos'. Caracas, 1986.
A.R. MARTINEZ *The journey from Petrolia.* Caracas: Petróleos de Venezuela, 1986.

A.R. MARTINEZ *Cronología del petróleo venezolano,* 6th ed. Caracas: Edreca, 1987.
H.M. VELARDE Ch. *La explotación unificada de los yacimientos de hidrocarburos y su proyección en América Latina.* Caracas: Corpoven, 1987.

EARLY REFERENCES TO VENEZUELAN OIL

GONZALO FERNANDEZ DE OVIEDO Y VALDES *Historia natural y general de las indias y tierra-firme del mar océano.* Sevilla, 1535 (1a. parte); 1540 (2da. parte).
FRANCISCO LOPEZ DE GOMARA *Historia general de las Indias.* Zaragoza, 1552.
RODRIGO DE ARGÜELLES and GASPAR DE PARRAGA *Descripción de la Ciudad de Nueva Zamora, su término y laguna de Maracaybo.* Archv. Gen. Indias. Indif. Gen., 1579.
JUAN DE CASTELLANOS *Elegías de varones ilustres de Indias.* Madrid, 1589.
ANTONIO DE HERRERA *Historia general de los hechos de los castellanos en las islas y tierra firme del mar océano.* Madrid, 1ª. Década, 1730. (Madrid: Primera, 1601–15.)
P. JOSEPH GUMILLA *Historia natural, civil y geográfica de las naciones situadas en las riveras del río Orinoco.* Barcelona, 1791. (Madrid: Primera, 1741.)
FRANÇOIS (RAYMOND JOSEPH) DEPONS *Voyage a la partie orientale de la Terre-Firme, dans l'Amérique Méridionale, fait pendant les années 1801, 1802, 1803 et 1804.* Paris, 1806.
J.J. DAUXION LAVAYSSE *Voyage aux Iles de Trinidad, de Tobago, de la Marguerite et dans diverses parties de Vénézuéla, dans l'Amérique Méridionale.* Paris, 1813.
ALEXANDER VON HUMBOLDT and AIME BONPLAND *Voyage aux régions équinoxiales du Nouveau Continent, fait en 1799, 1800, 1801, 1802, 1803 et 1804.* Parîs, 1816–31. (Also *Reise in die Aequinoctial—Gegenden des neuen Continents in den Jahren 1799, 1800, 1801, 1803 und 1804;* Stuttgart and Tübingen, 1815).
C.H. LYELL *Principles of Geology.* Vol. I, Part I, Cap. 3. London, 1830.
A. CODAZZI *Sumario de la geografía de Venezuela.* París, 1841.
M. BOUSSINGAULT *Viajes científicos a los Andes Ecuatoriales.* (Trad. Sp). Paris, 1849.

TECHNICAL REFERENCES

N. NUGENT *Account of the Pitch Lake of the Island of Trinidad.* Geo. Soc., Tr., Vol. I, London, 1811.
H. KARSTEN *Beitrag zur Kenntniss der Gesteine des nordlichen Venezuela.* Zeit. Deutschen Geol. Gesell., vol. 2, pp. 345–61, map. Berlín, 1850.

H. KARSTEN *Über die geognostichen Verhaltnisse des nordlichen Venezuela.* (Dr. C.J.B. Karsten). Archv. f. Min. Geogn., Berg. u. Hutten. Vol. 24, N°. 2, pp. 440–70; 1852.

H. KARSTEN *Geognostichen Bemerkungen über die Ungebungen von Maracaybo und über die Nordküste von Neu-Granada.* (Dr. C.J.B. Karsten). Archiv. f. Min. Geogn., Berg. u. Hutten. Vol. 25, N°. 2, pp. 567–73; 1853.

G.P. WALL *On the geology of a part of Venezuela and Trinidad.* Geol. Soc. London, Quart. Jour., Vol. 16, pp. 460–70, map; 1860.

W. BRICEÑO MENDEZ *Informe presentado al Poder Ejecutivo* (del Edo. Zulia) *sobre la exploración de la región carbonífera de Tulé y los depósitos de petróleo, betunes, asfaltos y carbón que contiene el Estado.* Rep. "El Zulia ilustrado". Nos 19, 20, 21, 22 and 23: Ed. Facsímil, Fund. Belloso, Maracaibo (1965); 1876.

C.H. BULLMAN *Mineral resources of Venezuela.* Eng. Min. Jour. (New York), Vol. 45, pp. 340–42; 1888.

W. SIEVERS *Die Cordillere von Mérida nebst Bemerkungen über das Karibische Gebirge.* (Prof. Penck) Geog. Abhand., Vol. III, No. 1, Wien (Holzel), 1888.

W. SIEVERS *Ein Schlammvulcan, Hervidero, in den Llanos von Maturín.* (Prof. Umlautt) Deutsche Rundschau, año 20, N°. 9, Wien (Hartleben), 1893.

E. FORTIN *Une plaine de bitume au Vénézuéla.* C.R. Soc. Geog. (París), Nos. 11 and 12, pp. 221–4, 1895.

H. EGGERS *Die Asphalt-Quellen am See von Maracaibo.* Deutsche Geogr. Blatt. (Bremen), Vol. 19, N°. 41, pp. 103–194, map; 1896.

E. CORTESE *Escursioni Geologische al Venezuela.* Soc. Geol. Italiana, Boll., Vol. XX, N°. 3, pp. 447–469; 1901.

H. HIRZEL *Erdöl und asphalt auf den Inseln Pedernales, Pesquero und del Plata in Venezuela.* Chem. Rev. über Fett u. Harz., Vol. 10, N°. 12, pp. 275–7; 1903.

L.V. DALTON *On the geology of Venezuela.* Geol. Mag. (London), Vol. 9, N°. 575, pp. 203–210; 1912.

CH. F.Z. CARACACRISTI *Magnesite on the Island of Margarita.* Eng. Min. J. (New York), 12.4.1919.

G.W. HALSE *Oil Fields of West Buchivacoa, Venezuela.* American Assoc. Pet. Geol., Bull., Vol. 31, N°. 8, pp. 1238–44; 1937.

J.H. REGAN *Notes on the Quiriquire Oil Fields, District Piar, State of Monagas.* Bol. Geol. Min. (Venezuela), Vol. 2, Nos 2–4, pp. 187–201; 1938.

H.D. HEDBERG *et al* *Oil Fields of Greater Oficina Area, Central Anzoátegui, Venezuela.* American Assoc. Pet. Geol., Bull., Vol. 31, N°. 12, pp. 2089–2169; 1947.

H.J. FUNKHOUSER *et al.* *Santa Ana, San Joaquín and Santa Rosa Oil Fields, Central Anzoátegui, Venezuela.* American Assoc. Pet. Geol., Bull., Vol. 32, N°. 10, pp. 1851–1908; 1948.

J.E. POGUE *El Petróleo en Venezuela.* Asoc. Venezolana Geol. Min. Pet., Bol., Vol. 1, N°. 1, pp. 85–145; 1949.

G. ZULOAGA *Venezuela.* In "World Geography of Petroleum". American Geog. Soc., Pub. N°. 31, pp. 49–79; 1950.
E. MONSALVE CASADO *La lección del petróleo.* Caracas, 1952.
J.M. PATTERSON and J.G. WILSON *Oil Fields of Mercedes Region, Venezuela.* American Assoc. Pet. Geol., Bull., Vol. 37, N°. 12, pp. 2705–733; 1953.
HNO. N. MARIA *Petróleo de Cubagua para su majestad la Reina.* Rev. "El Farol", CLXXVI, May–June, 1958.
J.M. SISO MARTINEZ *"Ciento cincuenta años de vida republicana"*, In *150 años de vida republicana.* Caracas: Ed. Pres. de la Republica, 1963.
A. KROBOTH *Historia de la Escuela de Geología, Minas y Metalurgia.* GEOS, N°. 10; Caracas, 1964.
D. BENDAHAN *La contratación colectiva en la industria venezolana del petróleo.* Rev. Col. Abogados Dto. Federal (Venezuela), N°. 129, pp. 203–226; 1965.
J. TELLO *Historia del Petróleo en Venezuela.* "El Farol": 218, pp. 6–9; 219, pp. 38–40; 220, pp. 36–41; 1966–67.
G.L. CARRERA *El tema del petróleo en la novela venezolana.* XIII Cong. Lit. iberoamericano. Caracas, 1967.
E. MONSALVE CASADO *Aspecto jurídico sobre el régimen de explotación de las minas. e influencia de la Ley correspondiente sobre el desarrollo minero nacional.* Rev. Fac. Derecho (U. Zulia, Maracaibo); Nos. 23, pp. 133–56; 1968.
R.J. ALMARZA *Cronología del Petróleo Venezolano.* II Jornadas Técn. Pet. (Maracaibo); 38 pp.; 1968.
H. PEÑALOZA *La Venezolanización Progresiva de la Industria Petrolera.* Pet. Mito Juan, N°. 1; 16 pp.; 1969.
J.J. NAVARRETE *Consideraciones sobre la reversión en materia petrolera.* U. Simón, Bolívar, 22 pp., 1970.
G. ZULOAGA *1930–1970: Las Inspectorías Técnicas de Hidrocarburos.* "El Farol": 234, 1970.
R. HURTADO IZQUIERDO *Naturaleza de las concesiones para explotar hidrocarburos.* In "Pro Venezuela y la nacionalización del petróleo", Asoc. Pro Venezuela, Caracas, 1975.
A. MOLINA *El Petróleo en el cine nacional.* "El Nacional", Caracas, 4.9.77.
A.R. MARTINEZ *The Orinoco Oil Belt, Venezuela.* J. Pet. Geol., Vol. 10, No. 2, London, 1987.

OTHER REFERENCES

BCV (Central Bank of Venezuela) *Memoria* (annual); *Memoria Especial 1961–1965; Actividades petroleras en. . .* (separatas Memoria).
CVP (Corporación Venezolana del Petróleo) *Reportes Anuales.*
MMH (Ministry Mines and Hydrocarbons) *Anuario petrolero y minero de Venezuela* (pub. 1949, 1956); *Memoria; Petróleo y otros datos estadísticos* (1958); *Carta Semanal*; *Datos generales. . .*

OPEC (Organization of the Petroleum Exporting Countries) *Background information, National Oil Energy Profiles, annual reports.* Geneva and Vienna.
AAPG (American Assoc. Pet. Geol) *Foreign exploration developments,* "*Venezuela*"; (exploration review issue).
AVGMP (Asociación Venezolana Geología, Minería y Petróleo) *Boletín Informativo.* (until 1971).
FUNRES (Fundación Rescate Histórico) *Documentos desclasificados del Departamento de Estado de los Estados Unidos,* 1910–1925.
Boletín de la Riqueza Pública (Est. R. Andueza Palacio, pub. 1891–1893).
Boletín, Sociedad Venezolana de Geólogos (from 1965).
Enciclopedia de Venezuela. 12 Vol. Barcelona, 1973.
Revista, Sociedad Venezolana de Ingenieros de Petróleo (from 1971).
El Farol (Creole Petroleum Corp.).
El Zulia Ilustrado (1888–1891). Ed. Facsímil, Fund. Belloso, Maracaibo, 1965.
Ministerio de Fomento *Revista de Fomento* (later *Revista de Hidrocarburos y Minas*) (1950–1957).
The Oil and Gas Journal (weekly); in particular, "World-wide oil" issue.
Petróleo y Tecnología, later *Petroleum* (bimonthly); Maracaibo.
Petróleo Interamericano, Tulsa, Oklahoma, USA.
Petroleum Press Service, London.
Platt's Oilgram News Service (daily); *New York.*
Venezuelan Economic Review (weekly); Caracas.
Venezuela Up-to date (monthly); Embajada de Venezuela, Washington.
VOSA (Venezuela Oil Scouting Assoc.). *Weekly Newsletter.*
El Nacional (Caracas, daily); (part, "Pág. de Hist.", 1981–2).
Panorama (daily); Maracaibo.
Energía y Petróleo (weekly); Caracas.

INDEX

For Maria Iracema
with every good wish
for a VERY HAPPY
HOLIDAY SEASON,

Sincerely,

John W. F. Dulles

Austin
7 December 1987